The Patrick Moore Practical Astronomy Series

For further volumes:
http://www.springer.com/series/3192

Visual Lunar and Planetary Astronomy

Paul G. Abel

 Springer

Paul G. Abel
Leicester, UK

ISSN 1431-9756
ISBN 978-1-4614-7018-2   ISBN 978-1-4614-7019-9 (eBook)
DOI 10.1007/978-1-4614-7019-9
Springer New York Heidelberg Dordrecht London

Library of Congress Control Number: 2013941661

Printed on acid-free paper

Springer is part of Springer Science+Business Media (www.springer.com)

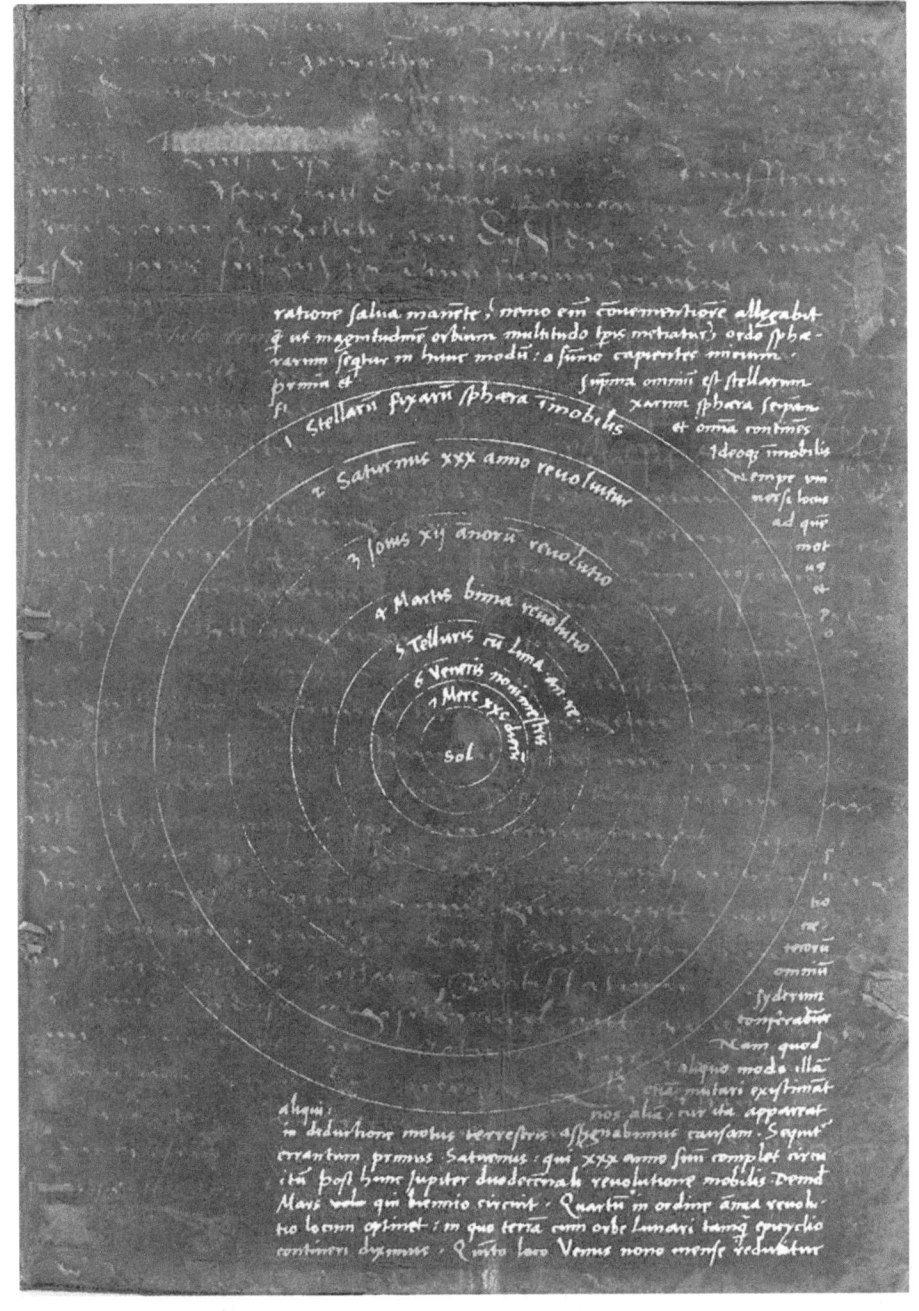

This illustration by James Westerman is based on a drawing of the Solar System done by Copernicus

Foreword

There was a time, decades ago now, when astronomers – both professional and amateur – used to depend upon drawing the objects under observation. Then came photography. Professionals soon depended entirely upon their cameras, and amateurs followed suit. There were a few professional lunar and planetary observers who continued to rely upon their artistic skill, but even in the amateur ranks these "sketchers" were rare, and when digital cameras replaced film only a few were left. The art did not vanish entirely, and there were a few who kept it alive, led by David Hardy and the late Paul Doherty, followed at a respectful distance by others (including me), but there were many observers who abandoned direct viewing entirely.

This was a pity in several ways. First, it is claimed that "the camera cannot lie." Wrong; it can, and an image easily be modified, either accidentally or by intent. Also, the human observer can take advantage of the few fleeting moments of perfect seeing conditions, which the camera cannot yet do so well. Thirdly, the experienced visual observer will be able to detect anything unusual, and lose no time in recording it.

Today there is a marked revival of sketching, particularly for Solar System work. It is not so easy as might be thought; materials have to be chosen and tested, techniques learned, and difficult problems solved. Accuracy is absolutely essential.

All this is why Paul Abel's book comes just at the right moment. Artistically he has now few equals and no superiors, and he is moreover a highly qualified professional astronomer at Leicester University, so that he is in an excellent position to advise and instruct. Read what he has to say, and you will realise that despite all the wonders of modern electronic techniques the rôle of the visual observer is as important as ever.

Sir Patrick Moore

Acknowledgments

I would like to acknowledge the following people for their help in preparing this book.

First to Sir Patrick Moore for helping me to become a consistent visual observer! I spent many years pouring over Patrick's log books, and my own are a template of his. I have his enthusiasm for visual work and his urging me to write this book to thank for bringing the book to life.

I must also extend great thanks to Richard Baum, a veteran visual observer and an excellent science historian. Richard kindly proofread every single chapter here and took the time to give me some excellent and honest feedback. Without his contribution I feel the book would have been another "standard text."

I would also like to thank Martin Mobberly, who went through the chapter on instruments and techniques, Dr. Matthew Forman for checking the chapter on the human visual system, and Professor Bill Leatherbarrow for reading through the Jupiter chapter.

My thanks must also go to the late Peter Hingley. Peter was the librarian of the Royal Astronomical Society and of great assistance to me when I needed access to observational materials and books from the late eighteenth and nineteenth centuries. Very sadly, Peter passed away in 2012. I am sure he had been a great asset to the RAS.

Finally, I would like to thank all those observers who contributed their wonderful drawings to this book: Sir Patrick Moore, Dr. Richard McKim, David Gray, Mario Frassati, and Dr. Carlos Hernandez. Their drawings have made the pages of this book come alive.

Many thanks also to James Westerman for producing the Frontispiece for this book.

About the Author

A lifelong amateur astronomer based in Leicester in the UK, Paul Abel earned his degree in Mathematics at the University of Leicester, specializing in general relativity and black holes during his final year. His Ph. D. From the University of Leicester is in Theoretical Physics and was concerned with Quantum Field Theory and Black Hole Thermodynamics. He is a co-presenter on the BBC's 'The Sky at Night' TV show, where he tries to promote scientific visual observation and general amateur astronomy. Paul is a Fellow of the Royal Astronomical Society (FRAS), and Assistant Director of Saturn section of the British Astronomical Association. He has published many articles on science and popular astronomy and is the author and co-author of a number of books on the subject. He is currently based in the centre for Interdisciplinary Science at the University of Leicester.

Contents

Chapter 1

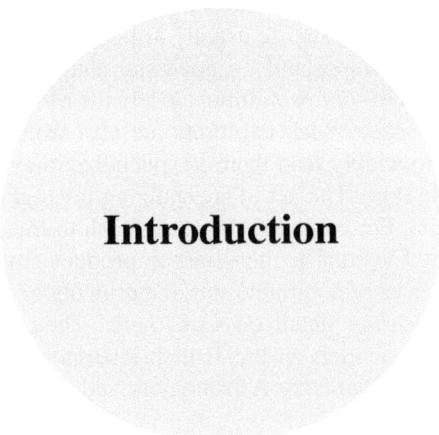

Introduction

We live in interesting times. There is an abundance of new technology, all of which quickly becomes affordable and accessible to the general populace. This advancing wave of technology is all pervasive, and it was only a matter of time before it reached the shores of amateur astronomy. This has been a mixed blessing.

The introduction of the CCD and webcam has allowed amateurs to produce great results that are of real use to professional astronomers, and this in itself has had a great warming influence, for it allows people who invest effort and time to get publishable results, one of the core purposes of amateur astronomy. In every way, the CCD has revolutionized amateur astronomy, but like all revolutions, there is a price to pay, and this price has been a subtle change in emphasis in astronomy, a subtle rewriting of its underlying principles.

The newcomer to astronomy today is likely to have been lured into the subject by fantastic pictures taken through modest- to large-sized apertures, and many have been led to believe that such images represent what can be seen through the eyepiece. A quick unpracticed, non dark-adapted view of such an object quickly leads to disillusionment. Moreover, the newcomer will probably be of the opinion that a vast telescope and computer are required. Very often beginners to amateur astronomy rush to install *Registack* and set about learning how to use the go-to function of their telescopes. In reality you should begin with the basics – learning your way around the sky and investing your time looking and becoming familiar with the various objects in the night's sky. Any other foundation is a foundation of quicksand. CCD imaging has perhaps made astronomy seem quicker and flashier and hides the amount of work that is needed to obtain good results with such equipment.

This is not an anti-technology book. Nor is this an anti-CCD book, for many of today's best imagers started off their careers as visual people. Indeed the idea of Imager vs. visual may seem like a real division, but it is in fact nonsense.

P.G. Abel, *Visual Lunar and Planetary Astronomy*, The Patrick Moore Practical Astronomy Series, DOI 10.1007/978-1-4614-7019-9_1, © Springer Science+Business Media New York 2013

The real division lies in the two general, much broader philosophies – the two different approaches to the subject. These are the methodical observer vs. the quick fixer. Let us look at these two mindsets.

The methodical observer starts life usually armed with a pair of binoculars or a small telescope, and most importantly, a good star chart. After the sky becomes a familiar place he or she will start examining closely the Moon and planets, and soon venture off into deep space. Such explorations, all chronicled lovingly in their observing log books invariably lead them to specialize; they may prefer planetary observation or variable stars. The act of specializing invariably facilitates the need for specialist equipment. For example, they may wish to image the planets or pursue visual photometry. Eventually, they start to produce really valuable observations, perhaps a light curve of a variable star, the drawing of a sudden Martian dust storm, or an image showing detail on Ganymede. Their observations are peer reviewed by organizations such as the British Astronomical Association or the Association of Lunar and Planetary Astronomers and are invariably published in journals.

By contrast, the quick fixer is like a butterfly, constantly flitting from one astronomical topic – an image of the Sun one day, a brief look at the Moon the next. Rarely are any observations submitted to any organizations. Such people tend to disappear from the astronomical scene quite quickly. One gains so much more from a subject by contributing to it in a regular and orderly fashion rather than a random grasping at topics. Since amateur astronomy has a basis in science, making headway in astronomy involves engaging with the scientific philosophy. This is not difficult, and you do not need to be a scientist to work in a scientific manner. Enthusiasm and dedication is more than enough.

So, what specifically is this book about, and why should it be of interest to you? The book is concerned only with visual lunar and planetary amateur astronomy as pursued from an amateur scientist's point of view, a book more for the budding methodical observer. It is my hope that this book will persuade you to try for yourself this particular method of observing the Moon and planets, and moreover, adopt the sound tried and tested principles, methods and practices that go with it. Before we outline how the book is structured, and how best to use it, let us first indulge in a little myth busting.

For example, some useful questions from people turn up time and again. These questions tend to be based on remarks people have heard.

First up, and by far the most frequent question: *Do you need artistic talent to be able to participate in visual astronomy?* The answer is an emphatic no! It is certainly the case that the visual planetary observer will draw what they see in the eyepiece; indeed it is common for two drawings to be made, a rough drawing and a neat copy. The process of making a drawing at the telescope is not an artistic one, rather it is draftsmanship. When creating a piece of art the artist is free to use license to represent an object in whatever manner and style they choose. They are representing their subject in an abstract subjective form.

Scientific drawing, or Draftsmanship as it's called, is altogether quite different. In this case, it is best to think of your planetary or lunar drawing as a technical drawing.

It is vital that you record what you see as objectively and as honestly as possible- it is perfectly acceptable to use artistic techniques, provided they are used in an draftsmanship manner, to make the drawing more accurate. The end result may look very colorful and detailed, but you should no more regard your drawings as works of art than you would, say, an architect's plans for a house, or a circuit diagram of a microchip. So then, no artistic talent really required, just patience and objectivity.

The next question frequently asked is: *Isn't drawing at the telescope much less accurate than imaging or photography?* This is an interesting question because it demonstrates a peculiar lack of faith in something that has served humankind for many millions of years – the human visual system. As we shall see in the next chapter, the human visual system is a brilliant piece of engineering. We depend on it to allow us to assess extremely dangerous situations, such as crossing a busy road. If it was in any way inaccurate we should not be here! It is true that modern cameras have a higher resolution, but when it comes to looking through a telescope, the human eye collects light in an active way and can make use of the changes in seeing; the eye is naturally fitted with adaptive optics. With practice and training the human visual system can become a very accurate piece of optical equipment.

You should also be warned not to expect too much from your first views of the planets. A quick glance at Jupiter in average seeing conditions cannot compete with decades of observing! It takes time and practice to be able to eke out all of the fine details on the Moon and planets. It's rather like training to run a marathon – you wouldn't on your first day of training, run the whole distance!

How to Use This Book

So, how best to use this book? After the introduction, the next section concerns the human visual system. This is an essential piece of equipment, and it's like your telescope inasmuch if you don't understand how it works and how to use it, you will never get the best out of it, for one of the secrets to becoming a reliable visual observer is learning the difference between looking and seeing. Most people don't realize that such a difference exists. A casual observer will look at Jupiter and see a few features. An experienced visual observer will see the subtleties in color, the faint storms and probably be able to detect their motion.

Next is a chapter on equipment, namely telescopes and eyepieces. It is not difficult to get advice on this subject. Everyone has an opinion on which type of telescope and which eyepiece and filter are best for one thing and another. To be honest, the best piece of equipment is the one that suits you, and to this end included here is a review of the main types of telescopes, eyepieces and filters. Also included are some details on how to calculate quantities such as magnification and some notes on how to keep an observing log book, for it is vital that you keep an accurate record if progression is to occur in any meaningful sense.

After this come the planetary and lunar chapters, Mercury, Venus, the Moon, Mars, Jupiter, Saturn, Uranus and Neptune. Now, rather than going through these in order, you should start with whatever is visible in the evening sky now! You will find the chapters contain the following sections.

Known Facts

Obviously we have so many facts concerning all of the planets to include them all would turn this book into a massive tome indeed. Therefore this section is restricted to the facts that are directly relevant when understanding what one is observing. As well as physical facts, also included are details about conjunctions and oppositions (superior planets) as well as how these objects may appear in various sized telescopes.

Historical Observations

If we do not know the history of a subject, we stand little chance of gaining an understanding of it and contributing to its future! Human beings have been studying the Moon and planets with telescopes for hundreds of years now, and there are many intriguing observations that have yet to be explained. For sure many can be put down to misaligned optics or bad seeing conditions, but as we shall see, other observations such as the ashen light and the blunting of Mercury's 'horns' have still observed in recent times. So as well as understanding the role visual astronomy has played in planetary science, it is recommended that you familiarize yourself with these oddities as well in the hope that we may one day have an explanation. This can only happen if reliable repeat observations are made.

Observing Details

In this section are details on how to observe and record the details of the Moon and planets. It should be stressed that the hints and tips given here are not just the author's but are based on the accumulated wisdom of Patrick Moore and many others fine visual observers who have all helped this author to become a reliable visual astronomer. As well as drawing techniques there are some practical notes on how to create correct observing blanks, calculation of phase (where appropriate) as well as examples of how to calculate longitudes and latitudes of various features on planetary disks. As we shall see, the free software *WINJUPOS* can be of invaluable assistance!

Observing Projects

In this section are details of regular observing projects that one can undertake. To be honest, these observing projects can be found all over the Internet, for many of them form the basis of the observing guidelines of the planetary sections of the British Astronomical Association (BAA) and the Association of Lunar and Planetary Astronomers (ALPO). However, it is useful to have them in one place, and included are details on making intensity estimates, transit timing and map making.

Finally, before we start, it should be said that in order to get the best out of your observations, you really should send them to organizations such as the BAA and ALPO. Not only will they put your observations to good scientific use, but you will also get objective feedback on your observations. This is vital if you are to improve; it's important to know what your strengths and weaknesses are so that you can improve upon them.

So then, let us now obtain some insight into this wonderful piece of optical equipment we are blessed to possess – the human visual system.

The Human Visual System

The human visual system (HVS) is one of the most remarkable systems ever to have evolved in nature, really every bit as wonderful and remarkable as the planets of the Solar System. It has taken millions of years of evolution to produce, and yet most people never give their visual system much thought unless it goes wrong. We all of us owe the HMV a very great deal; it has allowed humankind to make observations of the universe. These observations that have led to a better understanding of the physical world within which we live, and also to the development of science and engineering that have made all our lives better. As its importance is so intrinsic to visual astronomy – more so than telescopes and other optical equipment – we have included a short section on it here.

As you might expect, the HVS is a very complicated piece of equipment, and it is fair to say that there is still much to learn and understand. Indeed, research in this area is constantly progressing, and to give a detailed review of it here would not be practical! What we can do is to present a short discussion on the main components of the HVS and look at some topics that are immediately relevant to visual astronomy.

The Components of the Human Visual System

We start by taking a brief look at the components of the HVS, and the way these operate together to carry out its functions. Only the briefest of introductions is given here, and if you are interested in reading further, you might read *Foundations*

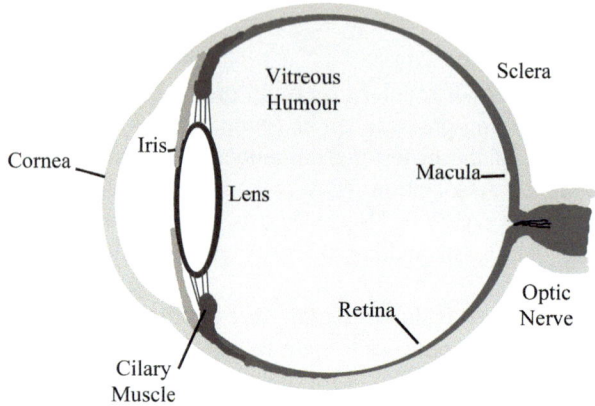

Fig. 1.1 Cross-section of the human eye

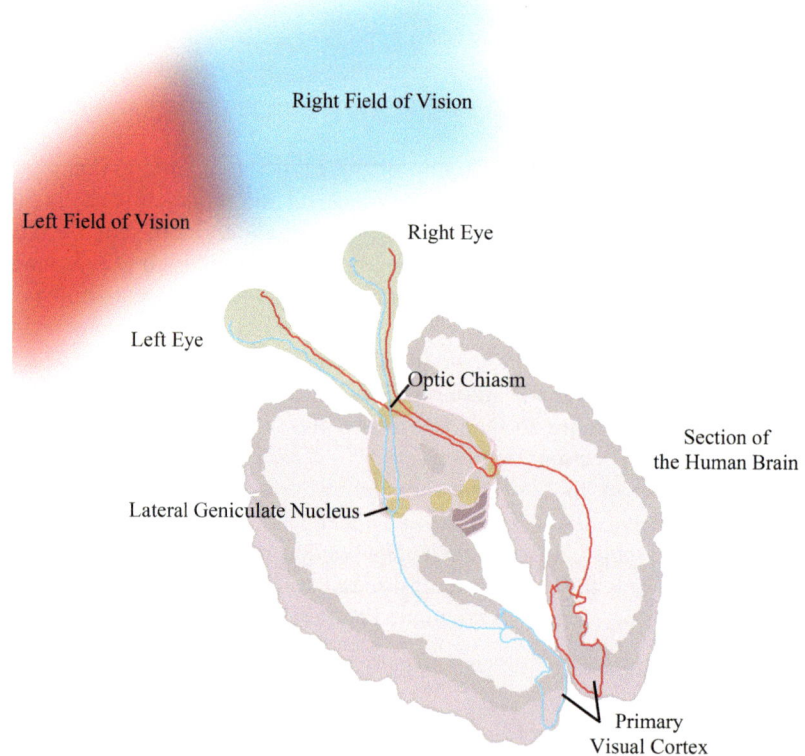

Fig. 1.2 The components of the human visual system

of Vision by Brian A. Wandell (1995). In Fig. 1.1 the basic structure of the human eye is shown, while Fig. 1.2 summarizes the components of the complete human visual system.

The best way to explain how all of this works is to look at what happens when we put our eyes to the eyepiece of a telescope. The light from the image in the eyepiece passes through the lens and is focused onto the retina, a layer of cells at the back of the eye. The retina is composed of two distinct types of cell:

- *Rods:* These cells are found around edges of the retina and are used mostly when light levels are low. Averted vision works because of these cells. Since rods are located around the periphery of the retina, if you look at a faint object (such as the faint cluster M.13 in Hercules) indirectly, you can see more details in the object than by directly looking at it. This is because the light is falling on the more sensitive rod cells.
- *Cones:* These cells are used to sense color. In fact there are three types of cells that respond to different wavelengths of light: short wavelengths cones (blue), medium cones (green) and long wavelength cones (red).

The chemical rhodopsin is manufactured in the retina. Rhodopsin allows us to see in the dark. It takes about 30 min or so for the chemical to be synthesized, which is why when you leave a bright room and step outside it takes some 30 min to become fully dark adapted. It takes your eyes that long to make rhodopsin. Alas, the chemical is rather light sensitive, and a quick flash of white light is enough to break it down. This is why astronomers use red light in their observatories. Rhodopsin is less sensitive to red light and hence this protects night vision.

The image from the eyepiece is focused onto the retina, which then converts it into a series of electrical impulses and passes it down the optical nerve to the optical chiasm. The optical chiasm is situated at the base of the hypothalamus (which is located in the brain). The optic nerves from both eyes meet here, and the signal is split into two – a left visual field and right visual field. The information from the right visual field (now on the left hand side) travels down the left optical tract, while information from the left visual field travels down the left optical tract. Both tracts meet in the lateral geniculate nucleus (LGN).

The LGN together with the visual cortex (located at the back of the brain) are responsible for image processing. The LGN sends its information to the visual cortex via cells in the optical radiation and the visual cortex, and the LGN can receive feedback from the visual cortex. The process whereby the brain decides what actions to take happen here with the interaction of the LGN and the visual cortex. This is why you will get the most out of your astronomical observing by treating it as a feedback process. The more you 'ask' your eyes what you are seeing, the more they will respond.

It is important to realize that the HVS is a feedback process because this observation leads to an important point. People often ask me, "how do you see the fine details on the planets?" After some discussion it normally is revealed that they have had a couple of looks at a planet for a few minutes and usually not seen a very great deal.

Fig. 1.3 The Rubin vase has more than one interpretation

This is perfectly natural; you should not expect to see much the first time, anymore than a marathon runner would start their initial training by running the whole distance!

The fact is, the more you go out and observe, the more you will see and the more responsive your eyes will become. In the end experience and telescope time are the keys to becoming a successful visual observer.

Perception and Objectivity

Perception is defined as the organization and identification of sensory information. Perception is probably one of the oldest fields related to the subject of experimental psychology, and it has taught us a great deal about how we perceive and understand the world around us.

Perception is not by any means a straightforward subject. How we perceive the world is not just down to the human visual system and biochemistry. Culture and other external factors are also involved. Moreover, perception is entirely subjective, and one object can be experienced in different ways by the same individual, for example the Rubin vase in Fig. 1.3.

The Rubin vase can either be viewed as a vase or as two faces (one face on the left and one on the right.). Both interpretations are valid and this has direct implications for visual astronomy. How are we to know what is real? The truth is, we can never be 100 % certain. There will always be some unreliability due to perception differences between observers, but if we are systematic this can be reduced; even imaging eventually requires human input in the post processing so there is no way to escape perception differences completely.

In a sense, visual observing is the constant management of self doubt; what we can do is train our visual systems to deliver the best objective results we are capable of. By making our observing sessions systematic, recording our results in a consistent and objective way according to good scientific principles, the visual observer can produce useful, reliable observations that should complement the results obtained by our colleagues in the imaging field. We are not in competition with imagers; we should aim to produce results that provide an alternative but complementary product to planetary observation.

A Few Words on the Perception of Color

It is appropriate at this point to say a few words about color. In the chapters that follow, we will be visiting worlds whose surfaces and clouds contain the most beautiful and delicate pastel shades. If you intend to make color drawings, or work with color in anyway, it is vitally important to realize that color perception is a subjective process, and so we should expect experience of color to vary from person to person.

Color perception is an entire research area in its own right, and I can only say a few brief words on it here. One explanation for the difference is the retinal cone cells. Each person will have a different number of each type and is therefore sensitive to different colors (for example, you may be more sensitive to the blue end of the spectrum and thus find it easy to see the dusky marking in the atmosphere of Venus). There are also the psychological aspects to consider. Interestingly, even where you grew up may have an effect on how you perceive color as well as your emotional responses to it.

So where does all of this lead us? It leads to the inescapable fact that each of us have our own biases and reactions to color, and we must be aware of this if we are to undertake useful scientific work where color is available. When we make color drawings we must be aware of the subjectivity introduced into them because of color, and the only valid and subjective way to measure the color of objects on planetary surfaces is to use filters whose transmission characteristics are documented and known.

We can get the best out of the human visual system if, like any tool, we know its capabilities, and its limitations and work with them accordingly.

Chapter 2

Instruments and Techniques

One of the hardest questions usually asked by newcomers who want to make a start with visual work is: *What type of telescope should I buy?* The question is difficult because the choice of telescope depends entirely on the circumstances of each individual. The best telescope for you is the one that suits your needs. The first thing to be done is to buy a pair of binoculars and learn your way around the sky. This is a major introductory step (and one that seems to be skipped all too often these days). It is true that many telescopes come with 'go-to' devices, but there is no substitute for familiarity with the sky.

Learning your way around the night sky will do two things for you. Firstly you will learn the constellations and become familiar with where things are situated. We all know how much quicker it is to get around a city if you know where all the back roads are. Yes you can use SatNav, but it is far more rewarding to *know* where things are. You are going to be spending a lot of time out there in the universe, so it is best to get acquainted with it. Secondly, being out there learning your way around the night sky will help make up your mind if you are going to be a causal amateur or someone who intends to invest a great deal of time in the subject. This is another consideration when buying a telescope; there is little point in buying a giant telescope and then having it sitting in doors collecting dust because it is too heavy to move about or because you haven't the time to use it.

For the rest of this book, I am going to assume that you have become familiar enough with the night sky and intend to make a proper start on visual observing. In the next section we shall look at the different types of telescope that are available along with the various choices of mounts, eyepieces and filters. After this we shall go on to look at some basic types of observatory that might be considered. In the final section of this chapter, we shall look at the tools and techniques needed to become a reliable observer.

P.G. Abel, *Visual Lunar and Planetary Astronomy*, The Patrick Moore Practical Astronomy Series, DOI 10.1007/978-1-4614-7019-9_2, © Springer Science+Business Media New York 2013

Although telescopes come in many shapes and sizes, it is possible to group them into three general categories – refractors, reflectors and catadioptrics. The term catadioptric can be confusing, but it simply implies a mixture of reflecting and refracting elements. The most popular modern form of catadioptric is the Schmidt-Cassegrain or SCT, but there are others, too, such as the Maksutov. Essentially, both these designs are similar to Cassegrain reflectors, with an extra transparent element designed to correct optical aberrations. It was only when Celestron's Thomas J. Johnson worked out how to mass produce large Schmidt corrector plates, in the late 1960s, that affordable SCTs became a reality for the amateur astronomer. Each types of telescope has its good and bad points, all of which must be weighed up before a telescope is purchased.

Refractors

This is the simplest form of telescope (see Fig. 2.1) and was the first type of telescope to be invented. It has a large primary objective convex lens (consisting of two or more elements) at one end whereby the light from a celestial body enters and is brought to a focus. This light forms an image at the focal plane that can be magnified and inspected by an eyepiece.

The main advantage of this telescope is that it is essentially maintenance-free. So long as the telescope is aligned carefully during manufacture and the lens cell is robust the refractor should remain in good alignment. Refractors also tend to have a large focal length (focal length is the distance from the primary lens to the point where the light is brought into focus, as is shown in Fig. 2.1). Longer focal lengths are useful because eyepieces perform better at long f-ratios and there is more eye relief, so spectacle wearers can observe more comfortably. It has to be said though, that refractors are very expensive telescopes. The smallest size of refractor useful for visual work is probably a 4-in. refractor, and this will cost more than a reflector of the same aperture. However, the real problem is if you want an aperture greater than 5 or 6 in.; at this point the cost of the instrument rises exponentially.

The main problem that affects the refractor is chromatic aberration. This occurs because white light is really a mixture of all possible colors from red through to blue. A glass lens refracts red light slightly less than it does blue light. The end result of this is a false color image; the object in the eyepiece will appear to have a purplish halo surrounding it. This was a big problem for the early refractors, but there have been many advances in optical production in recent years, and today, all refractors have a primary achromatic doublet objective lens. The doublet lens is essentially two lenses 'cemented' together, and the chromatic aberration of one lens cancels out that of the other, thus reducing the effect. All refractors in the medium price range will have a tolerable degree of chromatic aberration. There are a

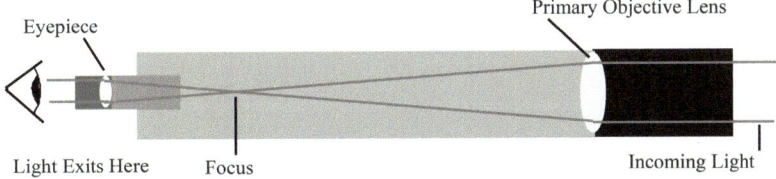

Fig. 2.1 A refracting telescope

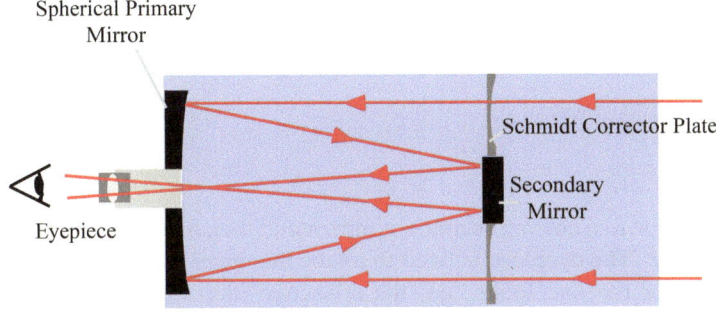

Fig. 2.2 A Schmidt-Cassegrain telescope

number of high end manufacturers who use exotic forms of glass in their refractors that eliminate the effect completely. However such telescopes are very expensive.

A further consideration is that the standard refractor requires a lot of room, and a 6-in. refractor is basically not portable. I was once loaned a 4.5 inch refractor by the British Astronomical Association back in the 1990s; it was an excellent instrument, but the optical tube was longer than me! These days refractors come with much shorter tubes, but at considerable cost and apertures of 5 in. or more, however the cost of smaller refractors has come down in recent years. Although they look splendid remember it is aperture (size of the telescope) that is the most important. Ideally, you should get the largest telescope you can for your money

Schmidt-Cassegrains

These types of telescopes are probably the most complicated for the beginner to understand and are based upon the principle of folded optics. A diagram showing how this type of telescope works is given in Fig. 2.2. Essentially, light from a celestial body enters the telescope through the corrector plate (which corrects for spherical aberration introduced by the primary spherical mirror); it then hits a spherical primary mirror at the back of the telescope. This light is then reflected

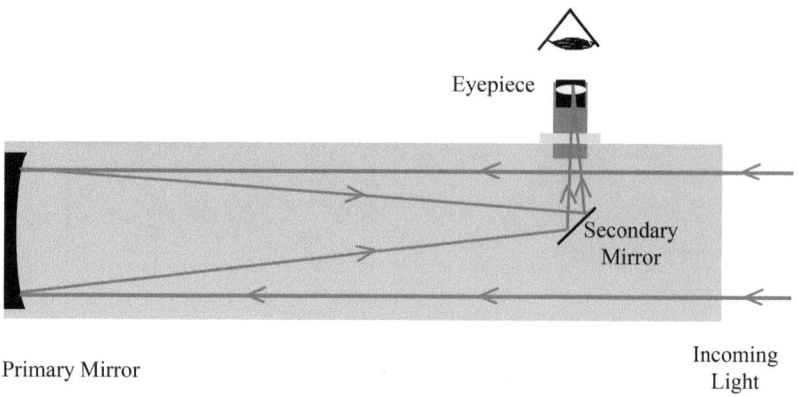

Eyepiece

Secondary
Mirror

Primary Mirror

Incoming
Light

Fig. 2.3 A Newtonian reflector

back up the tube to a convex secondary mirror and then back down through a tube to the eyepiece. The eyepiece is fixed to a tube that goes through the center of the primary mirror (so one looks 'up' the telescope like a refractor). The tube is sealed, and the telescope is brought into focus by moving the primary mirror back and forth.

The good thing about these telescopes is that they are very short. This means you can have high focal lengths, but the tubes are nowhere near as long as a refractor of equivalent focal length would be. This type of telescope does tend to be favored by imagers for this reason. There are however a few drawbacks. Schmidt-Cassegrains are rather expensive, they are also prone to losing their collimation (collimation means that the optics are correctly aligned). Moreover there is a loss of contrast due to the large secondary mirror diameter, which is far bigger than the hole in the primary.

Newtonian Reflector

The Newtonian reflector is a popular and reliable telescope. Money-wise they are cost effective, and most importantly you can obtain a large aperture telescope for a reasonable sum of money.

The Newtonian telescope is named after Sir Isaac Newton who designed the telescope and presented it to the Royal Society in 1668. The design is a simple one (see Fig. 2.3). The light from a celestial object enters the telescope tube and hits the large concave primary mirror at the bottom of the tube. The light is focused by the mirror and reflected back up the tube to a flat secondary mirror and is then reflected into the eyepiece. This type of telescope will need collimating from time to time (which is a simple enough process), but not as often as the Schmidt-Cassegrain.

As the tube is not sealed, the mirror will also have to be re-aluminized occasionally. There is zero chromatic aberration with the Newtonian telescope so color and filter work can be undertaken.

It is true the Newtonian's can come with long tubes if a longer focal length and high focal ratio is required (focal ratio is the focal length of the telescope divided by the diameter of the telescope and is written f/number). Although you will hear it said high focal lengths and ratios (example, f/7, f/e, etc.) are desirable for planetary work, telescopes with a focal ratio of f/5 can be very satisfactory. In end you must weigh up all of the factors and choose the telescope that really suits you and your needs.

The biggest advantage of having a long focal ratio Newtonian is that you can keep the collimation perfect over long periods of time. Newtonians suffer from an optical aberration called coma which increases exponentially as the f/ratio gets smaller. Not only does this lead to tadpole-shaped stars at the field edge, it means the region in the center of the drawtube, where the telescope image is perfect, can become very narrow. Independent of aperture this so-called 'sweet spot' is a blissful 11 mm across at f/8, a tolerable 5 mm across at f/6 and just over 1mm across at f/4. So, at f/4, if your drawtube rattles by more than 1 mm you are never collimated! Also, at f/6 and longer f-ratios, a telescope can be collimated in daytime, with a so-called Cheshire sighting eyepiece, without a star test being essential. The main disadvantage of a long f-ratio is that the telescope size can become very user-unfriendly.

There are many telescope manufacturers and suppliers today, and since a telescope is a large investment, it is strongly recommended you do some research first. Most (if not all) telescope makers have websites, and it is wise to read up about the construction and performance of a telescope. You might also find it useful to consult astronomy magazines and forums to get reviews from people who have used various brands of telescope. Your local astronomical society is another good place to go to. The members there will probably have many years of experience with telescopes, and you will be able to see the various sizes of instrument and use them for yourself. This will help you narrow down the choice of telescope that is suitable for you.

Mounts

"Your telescope may have the best optics built by human hand, but if the mount it is as steady as a jelly, it will be practically useless!" These were the words of wisdom from Sir Patrick Moore many years ago, and never was a truer word was spoken! The mount is, of course, the structure the telescope sits on, and it must be firm and rigid. If it is not, then the slightest breeze will make it shake, and your image of the planet will dart around in the eyepiece, making useful observation impossible.

There are two categories of telescope mount. The first type is an alt-azimuth, the second is an equatorial mount. We shall look at the alt-azimuth mount first.

Fig. 2.4 A Dobsonian mount

Alt-Azimuth Mountings

This is the simplest type of mount a telescope can have. It generally comes in the form of a tripod, and the telescope moves in altitude (up and down) and azimuth (left to right). This type of mount is very easy to set up; you simply find some firm level ground, set the telescope down and you're ready to go. A good example of this type of mount is the Dobsonian mount (see Fig. 2.4). The Dobsonian mount is very popular with deep sky enthusiasts. Firstly, this type of mount doesn't take up too much room and so quite large telescopes can be mounted in this way (it is not uncommon to see 20″ Dobsonians at star parties). The Dobsonian is also very quick and easy to set up, and you can move from one part of the sky to the other very quickly.

The main drawback for this type of mount is the fact that it is not usually driven and therefore does not follow Earth's rotation (drives are available for Dobsonian mounts, but they are very expensive). All celestial bodies in the sky gradually move through the sky during the course of the night; as Earth rotates, objects rise in the east, culminate (reach their highest point) and then set in the west (unless they

Fig. 2.5 Equatorial mount. The mount is set up so that the polar axis is pointing in the same direction as the North Pole (or South Pole if you live in the southern hemisphere). The counter weights are there to make sure the telescope is balanced

are circumpolar, which means they never rise or set and are visible all year around, like the plough in the UK for example). As a result, when you turn the telescope towards an object (Saturn, say) you will find the planet slowly drifts out of the field of view of the eyepiece.

If you use a high power, the speed which this happens can be rather quick, and you may find yourself only getting a 20-s view! Slow motion controls can be added so that you can keep moving the telescope in small amounts, but to do this and make a drawing requires more hands than the human form is equipped with! If you have a small garden and want a big telescope, and don't mind frequently shifting the telescope a little, this maybe a suitable telescope mount for you.

Equatorial Mounting

This type of mount is more complicated (and more expensive). It moves the telescope in two directions, right ascension and declination. All celestial objects have coordinates measured in terms of right ascension (denoted α) and declination (denoted δ), and this is the coordinate system of the sky used by all professional astronomers (and indeed, go-to telescopes). As can be seen in Fig. 2.5, the polar axis of an equatorial mounting points in the direction of the pole star (and so is aligned with the rotational axis of Earth). As a result when the telescope moves in right ascension it is moving in the same plane as Earth's equator, and so with a motor driving this axis at a rate of one revolution every 23 h, 56 min and 4 s objects will stay in the field of an eyepiece.

Before we discuss the equatorial mounting further, it will be useful to remind ourselves on what is meant by right ascension and declination. Of course, you don't need to know anything about right ascension and declination to use an equatorial mount; to use it effectively you simply need to align it with the pole star. The main benefit is that a drive can be fitted so that the telescope rotates at the same rate as Earth (i.e., it keeps sidereal time), so once you have locked on to an object, it will remain in the eyepiece rather than drift off after a few seconds. Having said that, understanding the coordinate system used by astronomers is a great advantage, and since it is a relatively simple idea, we shall briefly look at how it works.

Imagine (or rather, pretend) that the whole night sky, including the stars, lie on a sphere surrounding Earth (completely unrealistic, of course, but for the point of choosing a coordinate system this doesn't matter!). This sphere is called the celestial sphere. We can now put some useful things on the sphere (see Fig. 2.6a). First we can put on the celestial equator, as well as the north celestial pole and the south celestial pole. We measure right ascension in the horizontal direction (so it is equivalent to longitude here on Earth), and we measure declination in the vertical direction (and so declination is equivalent to terrestrial latitude).

Declination is the simplest coordinate of the two. Declination is measured in two directions – up (north) and south (down). It is measured in degrees (°), minutes (′) and seconds (″). Each degree is composed of 60 min, and each minute is further composed of 60 s. Anything on the celestial equator has a declination of 0°. Everything north of the celestial equator will fall into the range 0° (equator) to the north pole located at +90°. In the other direction (below the equator), an object can have a declination ranging from 0° to −90° (the south celestial pole). Thus a ' + ' indicates that the object is above the celestial equator, while a '−' indicates it is below the celestial equator. There will be objects in the southern hemisphere that cannot be seen in the north (and vice versa). For example, astronomers in the UK and northern Canada will never be able to see the magnificent constellation of Centaurus simply because it is too far south and never rises above the horizon (thus we are denied the splendid sight of the magnificent globular cluster Omega Centauri).

Right ascension is a circle in the horizontal direction on the sphere, and it's split into divisions whose chosen units are hours (h), minutes (m) and seconds (s). One complete circle is 24 h in length; each hour contains 60 min, and each minute is broken down into 60 s. The circle starts at 0 h, half the circle is covered after 12 h, and after a full 24 h we are back at the start of the circle. On the right ascension circle, we start at 0h and move east (to the right, in Fig. 2.6b) 1 h, 2 h and so on. We can specify the right ascension of any object on the celestial sphere, with its coordinates given in hours, minutes and seconds (for example, Vega is located at about 18 h 36 m 56 s). As a coordinate system this is simple enough, but we have to decide where we put the 0 h.

The ecliptic is the path the Sun makes across the sky over the course of a year. In Fig. 2.6b we have included the ecliptic and the celestial equator. As can be seen,

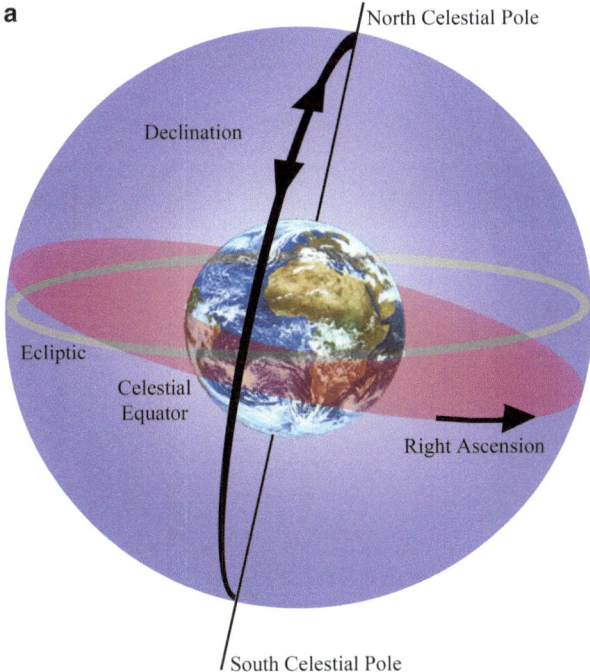

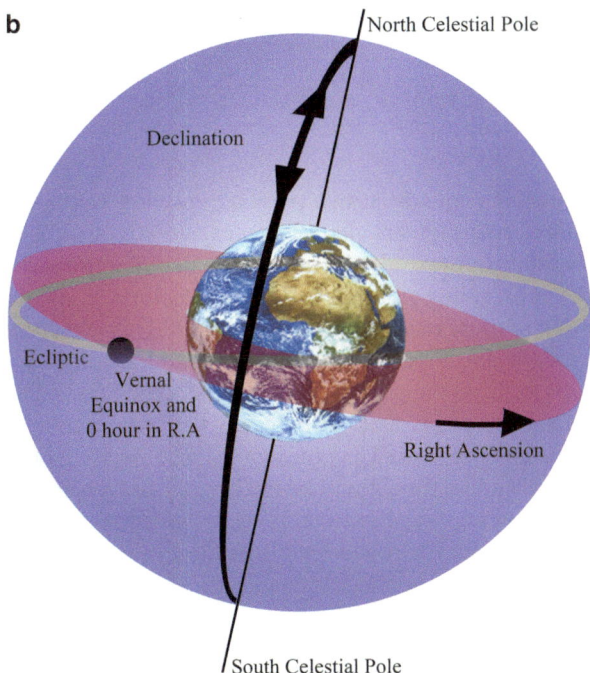

Fig. 2.6 (a) The celestial sphere. The celestial equator, poles and RA and dec are shown. (b) The ecliptic and equator intersect at the equinoxes. 0 is the point of the vernal equinox. (c) Vega's location on the celestial sphere

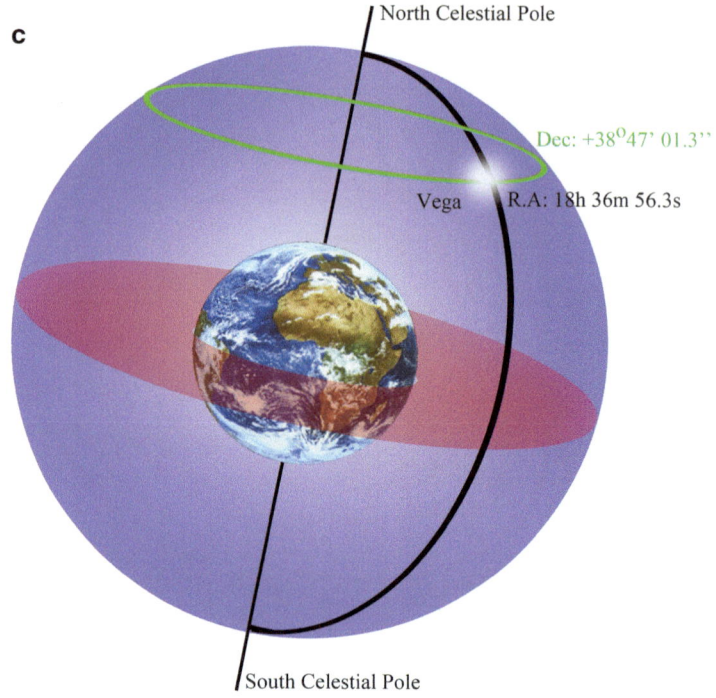

Fig. 2.6 (continued)

the ecliptic intersects the equator at two points, the spring equinox and the vernal (or autumn) equinox. Astronomers choose to measure 0 h of right ascension to be the point of the vernal equinox (also known as the first point of Aries).

Very often in books giving right ascension and declination of objects, you will see 'Epoch 1900' or in more recent texts 'Epoch 2000.' This is there to take account of the effect of precession. Precession occurs because Earth isn't a perfect sphere (it bulges at the center), and the gravitational effects of the Sun and the Moon cause Earth to wobble slightly on its axis. This means that the current North Pole star (which at the moment is Polaris in Ursa Minor) will not always be the pole star, due to precession; in 8,000 years time it will be the bright star Deneb in Cygnus, and in 12,000 years' time it will be Vega in Lyra. Polaris will return to being the pole star in around 26,000 years, which is the length of the cycle. This obviously has an effect on where the equinoxes occur, and due to precession, they shift slightly, even over the course of a century. As a result, right ascension has to be regularly updated for every object (once every 50 years is enough), and this is what is meant by Epoch. Epoch 2000 means the right ascension coordinates are given from the position of the equinox in the year 2000.

Now, what does all this have to do with telescope mounts? The answer is: everything! The equatorial mount is designed to make full use of the right ascension and declination coordinates. On simple 'non go-to' mounts you should find a setting

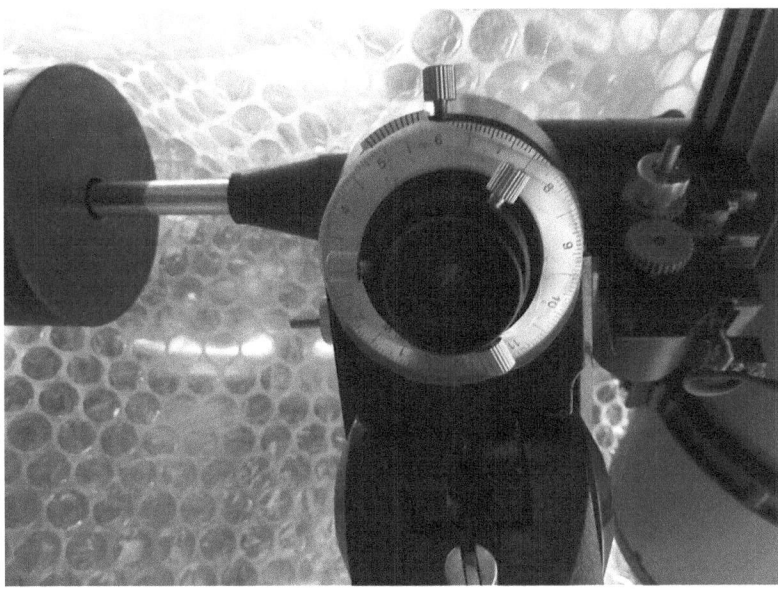

Fig. 2.7 A polar telescope. Simply look through it and adjust the mount until Polaris enters the polar 'scope. Once you have it, place it in the desired position (as indicated in your telescope's manual)

circle for right ascension axis that measures right ascension in hours, minutes and seconds, while on the declination axis you will hopefully find a declination circle to measure declination in degrees, minutes and seconds. If you know the coordinates (in RA and dec) of a faint object such as Uranus you can turn the telescope to these coordinates straightaway (hence having to avoid a search of an obscure area of the sky with no easily recognizable stars!).

To set up the equatorial mount, first make sure you have read the instructions that came with the telescope. Next, the polar axis needs to be pointing towards Polaris. Most equatorial mounts today have a polar telescope (Fig. 2.7). This is a small telescope located in the polar axis of the mount. You simply point your mount towards the north and in the direction of Polaris, and when it appears in the polar 'scope you move into position. (Polaris is not quite at the north celestial pole, so most polar 'scopes indicate the position where you should put the pole star once you have it. In some models there is a black circle in the polar telescope, and the pole star has to rest anywhere on that, depending on the month and time.) The telescope is now ready to use. If it is motor driven you will notice that once the drive is on, any object placed in the eyepiece field will stay there because the telescope is tracking it.

If you wish to find an object by using its RA and dec, then first find a bright star and turn the telescope onto that (Vega is a good choice in summer, Sirius in winter). Turn the RA circle so it reads the RA coordinates of the star, then turn the dec circle so it now reads the dec of the star. (The dec circle may be fairly close to the correct declination as delivered.) All you have to do then is to move the telescope to the RA and dec of the object you wish to observe. If your telescope is driven, you may only have to do this once per night, assuming the drive is not switched off. However, life is not always that simple! If you move to another object by using the fast slew button on the handset you may find that the setting circle RA position does not change at all, so studying how the RA circle can be clamped or released from the polar axis, and watching what happens will help understand its operation. Sadly, many modern telescopes are not designed with the setting circle user in mind and some simply use go-to, which can require a tedious multi-star alignment procedure. However, if the telescope is fixed on a plinth in an observatory you will never have to alter the declination setting circle and that, in itself, is a huge advantage, especially when searching for objects in twilight.

For many, the equatorial mount is ideal for visual observing. It is true that it can take about 15 min to set (make sure the telescope and the counter weights balance, and it is aligned with Polaris), but the advantages of having a planet steady in the eyepiece without any drift far outweighs the work in setting it up. All of your energy can be spent observing and recording, rather than having to make adjustments to the telescope every minute or so.

Computer Drives

The option exists today for a telescope on any mount to be controlled by a computer drive. The difference between a clock drive and a computer drive is that a clock drive will simply turn the telescope at the same rate as Earth, whereas a computer drive has a go-to system, and so all you have to do (in theory) is connect the telescope to the computer and tell it where you want to look.

Go-to systems are not too difficult to set up, but it has to be said they do drive up the cost of a telescope; so much so that even a small telescope on such a mount is bound to be costly. These drives are ideal for deep sky enthusiasts who wish to track down faint elusive objects, and photograph them for long periods of time, but planets are bright and easy to spot, so one has to question whether much use will be gotten out of them for the planetary observer. The would-be telescope buyer might be wise to invest the money in a bigger aperture (since this is more important) rather than a computer-driven telescope mount. However, in general the mounts that can cope with instruments of a 10- or 12-in. aperture do tend to have 'go to' as standard these days, and so many planetary observers will simply bypass the date, time and multi-star alignment menu routines, go straight to the sidereal (or lunar) tracking command and just align the 'scope very roughly on the pole. Even crude marks on the patio telling you where the tripod feet should go will suffice for a quick set-up and, in the cool, overcast climate, a quick set-up is often vital to beat the clouds!

Eyepieces and Filters

Once you have a telescope, a selection of eyepieces is essential. To start off with, it's best to have three eyepieces, a low power one for hunting down faint planets including Uranus and Neptune, a medium power one that will provide reasonable views in medium seeing, and finally a high power one that will give close up views of the Moon and planets when the seeing is good. It should be stressed that even the lowest power eyepiece is no substitute for a decent finder telescope and a high quality finder makes life much easier. Many experienced amateurs have two finders, a zero power laser-type finder and one with an aperture of 50 or 60 mm. Such an arrangement is a 1,000 times more useful than a small 'toy' finder of 20 or 30 mm aperture.

Before we look at the different eyepiece types, let's review magnification and how to work it out. Simply put, magnification is the amount by which an eyepieces increases the *apparent* diameter of an object. So if your eyepiece is ×50, then when you use it on the Moon, the Moon will appear 50× larger than when you see it with the naked eye. Magnification depends on two quantities: the focal length of the eyepiece f_E (see Fig. 2.8) and the focal length of the telescope f_T, and they are related to magnification by the expression:

$$magnification = \frac{focal\,length\,of\,telescope}{focal\,length\,of\,eyepiece} = \frac{f_T}{f_E}.$$

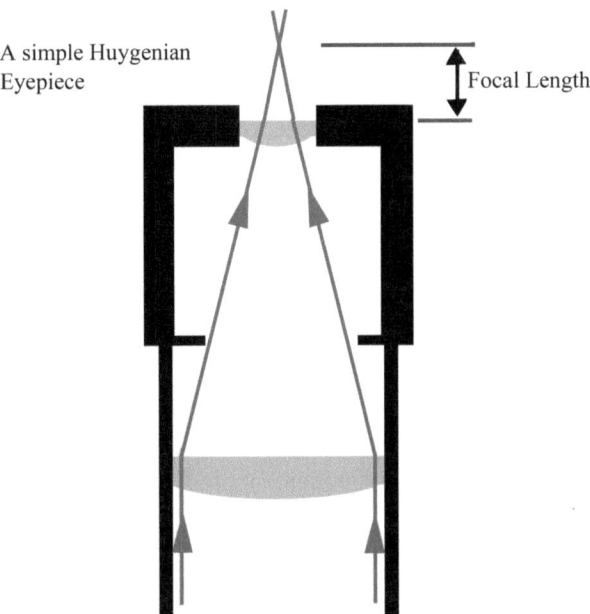

A simple Huygenian
Eyepiece

Focal Length

Fig. 2.8 Focal length of eyepiece

Table 2.1 This shows the maximum usable magnification for various-sized telescopes. Values are approximated for convenience

Telescope size	Maximum usable magnification
3 in. (76 mm)	152×
4 in. (102 mm)	204×
5 in. (127 mm)	254×
6 in. (150 mm)	300×
8 in. (203 mm)	400×
9 in. (230 mm)	460×
10 in. (250 mm)	500×
12 in. (305 mm)	610×
16 in. (406 mm)	812×

The focal length of your telescope will be given with its packaging and instructions. My telescope is of focal length 1,000 mm, so my 5 mm eyepiece will give a power of 200× since

$$magnification = \frac{1,000}{5} = 200$$

However on a telescope with a focal length of 1,500 mm we find that

$$magnification = \frac{1,500}{5} = 300.$$

So, you need to remember that if you use your eyepieces on a different telescope, unless the focal length is the same as your own telescope, the magnification will be different. Note also, both the focal length of eyepiece and focal length of telescope must be in the same units (mm is the most common).

Magnification is also dependent on the size of your telescope. You will find that as you increase the power, the image becomes dimmer and harder to focus. If you have a 150 mm reflector, you will find that an eyepiece that gives a power of ×500 gives an image that is practically useless! There is a maximum usable magnification for each telescope, and this maximum will only be achieved on very good nights with excellent seeing (Table 2.1).

The general rule of thumb for working out the optimum magnification is to take the diameter of your telescope (in mm) and multiply it by 2. Thus a 14-in. (356 mm) telescope will have an optimum magnification of

$$356 \times 2 = 712 \text{ times.}$$

Seeing conditions will also affect what magnification you can use. One scale used by astronomers to assess the seeing is the scale introduced by Eugene Antoniadi. The scale ranges from Ant I (Perfect) to Ant V (Unusable):

Ant I: Perfect seeing, image rocky steady with no turbulence and undulations. High powers of magnification can be used for long periods of time.

Ant II: Good seeing, some quivering in the image and occasional loss of image definition, but medium to high powers can be used without too much difficulty.

Ant III: Average seeing, almost equal moments of good definition and poor definition. High powers can be used occasionally.

Ant IV: Poor seeing. Some steady moments, but largely the image frequently quivers and blurs. Image definition may be poor even in medium powers. All fine details are lost, very little can be done.

Ant V: Bad seeing. Nothing can be done at all, the image 'boils' away in the eyepiece, and it is impossible to focus; very few (if any details) seen on the object in the eyepiece.

Only in the good seeing of Ant II or more will you be able to use a high power on your telescope, and in some cases you may be able to exceed the maximum usable magnification.

Choosing an Eyepiece

Nearly all of the standard manufacturers of telescopes include eyepieces with the telescope, but it is always worth buying a few extra. In times gone by there were many different types of eyepiece available, and some of them were very poor. Today, nearly all eyepieces are of a good quality and can be purchased at a reasonable price. Cost is likely to be the most significant factor, followed closely by eye relief (the distance the eye needs to be from the eyepiece). You might want to have a range of eyepieces, from the cheapest to the most expensive such as a wide field planetary eyepiece (see Fig. 2.9). Certainly one can spend a lot.

A wide field eyepiece allows you to see more of the field surrounding an object. This is ideal if you're using a medium power and examining the galaxies of the Virgo cluster, but there is no great advantage in having a wide field eyepiece for planetary use (unless of course you wish to see all of the moons of Saturn in the field). The difference between the *real* field of an eyepiece and its *apparent* field can sometimes confuse the beginner, but it is best explained with an example. The Moon is roughly half a degree across, as seen with the naked eye. Imagine you are looking through a telescope at the full Moon, and it fills the eyepiece perfectly, at a magnification of 120×. The Moon is still half a degree across, which means that the real field of the eyepiece/telescope system is half a degree. However, the apparent field of the eyepiece must be 60°, because even at 120× magnification a half degree object can still fit into the field. Some enormous apparent field eyepieces are now available if you have deep enough pockets. For example, the TeleVue Ethos range now allows apparent fields of 100°, so that the Moon would still fit in the field

Fig. 2.9 A selection of eyepieces

at 200×! The view is literally like floating in space, but the cost of each eyepiece is enough to buy an entire telescope.

When it comes to selecting an eyepiece, start by looking at the websites of various telescope dealers. Very often you will find the firm that manufactures eyepieces has a website, and you can read all about a particular eyepiece that interests you. It is wise to do this preliminary research, as you can find out useful information about eye relief, or what the eyepiece is best suited for.

Barlow Lenses

A Barlow lens is an extra lens that boosts the magnification. It is fitted between the drawtube and an eyepiece. A common Barlow lens may be a ×2 (though ×3 are becoming increasingly common), and it effectively doubles the magnification of an eyepiece. Barlow lenses are very useful and well worth investing in; if you have three eyepieces and a ×2, then effectively you have six possible choices of magnification. They enable medium focal length eyepieces to be used at high powers, which usually gives a far more comfortable eye position for the observer. Beyond Barlows are the range of TeleVue Powermates, which allow a ×5 increase in effective focal length. These are especially useful to planetary imagers, as they enable a Newtonian of f/5 to perform like an f/25 system, enlarging the disk of a planet so it almost fills the imaging CCD.

Filters

A visual filter is normally a piece of glass (or plastic) coated in a material designed to block certain wavelengths and allow others to pass. Filters can be screwed into the eyepiece or used in a filter wheel, and they permit the study of objects in certain parts of the visible EM spectrum. This has numerous advantages for visual work; a blue filter will help enhance the white clouds of Mars, while a red filter will help to bring out the faint festoons of Jupiter's turbulent equatorial zone.

As we shall see in the subsequent planetary chapters there is a lot one can do with filters, and filter work will form part of the general investigative work that the visual amateur can undertake. Some telescope manufacturers include a set of filters with the eyepieces they provide, and the filters themselves are easily applied; one simply screws them into the base of the eyepiece (Fig. 2.10).

Filters can be found from many telescope manufacturers and dealers, and a simple online search will give the names and details of suppliers. You will find that each filter has a unique Wratten (W) number. The Wratten number was named for Frederick Wratten, a British inventor who, along with C. E. K. Mees, had a business that developed the range of optical color filters that they then sold to Kodak in 1912. The W is written on the side of the filter, and this corresponds to the color of the filter and the wavelengths it transmits. Filters really extend what the visual amateur can do. Throughout this book we will be referring to filters by their Wratten number and their colors.

Fig. 2.10 A selection of some filters; the W# is written on the side

An observatory is not an essential requirement for the serious amateur, but it does make life considerably easier! Part of the problem with heavy mounts and large telescopes is that it can take some effort to haul the telescope and equipment outside and get it set up (especially if polar alignment is required), by which time the clouds have swamped the night sky. This can not only be disheartening but actually off-putting. A number of people have told me how their astronomy interests began to wane after several such occasions. You can exploit even half an hour of clear sky if you're ready to go in a few minutes.

Traditionally, an observatory takes the form of a large white dome. The telescope is enclosed within, and a shutter in the dome is opened when the telescope is to be used. Although one can purchase domes of various sizes for amateur telescopes, it is not the only option available. If you decide to have an observatory, the first thing it must have is a stable floor. If the floor is in the least bit unsteady, every minor tremor will be picked up by the telescope. The standard method with a dome is to have the telescope pier completely isolated from the floor, so observers moving about don't affect the instrument. Normally you have a concrete pier, set deep in the ground, and it comes up through a hole in the floor. A wooden floor is best as it does not store megawatts of heat, which then happily destroys the seeing all night long!

Dome seeing can be a problem in summer time. Heat can collect in the dome, and when you come to observe you find the air is as steady as jelly. If you intend to get a dome, make sure you open it a few hours before you plan to observe.

Today, there are many suppliers of readymade observatories, and if you have the funds available, then it is certainly worth considering getting one that is readymade. If on the other hand this is out of the question then don't despair; you can join the many amateurs who made their own observatories by creating or modifying existing garden structures.

As you might imagine, there are many different types of observatory one could create, and many people have come up with ingenious creations. Indeed there are now books devoted to the subject. With that in mind, we shall just go through some of the common types that are easy to assemble. The simplest type is known as a run-off shed (see Fig. 2.11). A run off shed is composed of one or two parts that sit on rails and enclose the telescope when not in use. When the telescope is needed, the covering section is simply pulled back, and the telescope is ready to use. Although it is beyond this author's carpentry skills (which are incredibly poor!) such a shed could be constructed out of hardboard quite easily and cheaply. A set of wheels and rails would be needed to make sure the shed runs smoothly along the ground. Anyone who is particularly good with their hands might be able to modify some of the smaller plastic garden sheds for such a purpose. This type of observatory has the advantage that the telescope can be accessed quickly, although one is still exposed to the elements!

a

b

Fig. 2.11 (a) Sir Patrick Moore's 12.5-in. reflector. Moore's run-off shed was composed of two wooden sections. The two sections enclose the telescope and sit on rails. You simply pull them apart when you want to use the telescope. (b) Martin Mobberley's run off shed. Mobberley's run-off shed is a more modern system based on a plastic garden shed where the floor has been modified and six wheels added

Another option to explore is the run-off roof or fold back roof design (see Figs. 2.12 and 2.13). Essentially it is a normal garden shed, but the roof has been replaced with a lightweight plastic object. The roof is hinged and folds back when the telescope is in use. In other cases, the roof rolls back on wheels instead. Either way it is a good system, and one that has the added bonus of providing insulation against the wind. The observatory cannot of course be heated, since the heat would destroy the seeing, but it feels a good deal more sheltered than just being out in the

Fig. 2.12 The observatory that houses the author's 203 mm (8-in.) reflector

garden. You can add a small desk and some shelves to store equipment, and this is another bonus of having an observatory with everything all in one place. Run-off roof buildings are often used for Schmidt-Cassegrains or refractors mounted on a high pier, as sometimes using them with Newtonians can block a lot of the sky at low altitudes, which is of little use, for example, to the comet observer.

Drawing Equipment

Once you have a telescope, eyepieces and filters (and perhaps an observatory), the next thing to invest in is some drawing equipment. The first thing to get is a set of quality pencils that range from HB to 6B. Some of the heavier pencils (like 4B, 5B and 6B) are very good for dark markings and can be smudged very easily. Good quality drawing paper is essential, and it should be quite thick so that it can cope with various heavy pencils and ink. You will find it far easier to print planetary blanks onto your paper, rather than draw them free hand. A planet such as Saturn has at least six different ellipses to be drawn accurately! You can get planetary blanks from the British Astronomical Association website, and you should print all blanks onto good quality paper. It is worth experimenting with your printer to determine the thickest paper it will print on. Use the thickest paper possible so it absorbs the inks and water without distorting the paper. You may find it

Fig. 2.13 Gary Poyner's observatory

best to start making your drawings in black and white so you can experiment with various styles and techniques.

If you do decide to go to color, permanent water color crayons can be very good. You will need to experiment with them to get the correct tones (again this is why thick paper is essential as very often a dab of water is required to blend in the colors). You might also find Indian ink very useful. You can use this on lunar drawings for shadows, along with the shadows of the Jovian moons. You can find all of these sorts of materials at a local art shop. It is well worth shopping around, however, as you will find some stores charge substantially more than others. Since this is a recurring cost over time, it is good to know where the reasonable suppliers are!

Some other equipment that you may find useful includes a clipboard (useful when drawing or making notes), a red light and a good white torch as packing stuff away in white light seems to be far easier than in red!

Observing Practices and Record Keeping

It cannot be stressed enough how important it is to keep an accurate record of your observations. Not only will your log book provide a useful commentary and demonstrate how your drawing and observing skills have advanced as time progresses, they will also become scientific documents in their own right. They will be your scientific records of what was observed on a particular night and just as importantly, what was *not* seen. If someone wishes to check a specific planetary feature on a given date and time, you will be able to go straight to it if you have recorded your observation in a log book.

You might have several log books. First, there is the observatory log book, taken out to the telescope every time you observe, and into it you can put all of your rough observations, notes and drawings (along with notes of any meteors or fireballs that may happen in the night) during the course of a night. Then you might have separate notebooks inside for neat drawings. You might want to keep things separate, so you might have a book for Venus, Mars, Jupiter, Saturn and Uranus, as well as books for the Moon and the Sun. Into these books you can put the neat color drawings and the pertinent notes from each observation. Make your drawings on separate pieces of good quality drawing paper and stick them into the books after they are finished.

Your log books need to be sturdy, as they are going to be used quite frequently, and it is worth spending a little money on them so they don't fall apart.

Whether you decide to keep your observations in one book or separate books, it is a good idea to have a separate book for your finalized observations. One thing that should be mentioned is the temptation to use ring binders and loose leaf A4 paper to record and store your observations. This should be avoided since it's only a matter of time before the ring binders open accidentally and a decade's worth of observations become mixed up in a matter of moments!

It's a matter of personal choice how much you record in your log book, but here are some details that must be included if the observation is to have any merit:

- *Date*: day, month and year when the observation was made.
- *Time*: This should be Universal Time (UT), which means in summer time in the UK you will have to remove an hour.
- *Telescopes, eyepieces and filters*: those used to make the observation.
- *Seeing conditions*: You should record the seeing conditions using the Antoniadi scale mentioned earlier where Ant I is perfect seeing and Ant V is very poor.
- *Weather conditions*: It is always useful to record what the weather conditions are like during the time you are observing, for example if there is low scattered cloud or if it is windy, as this might affect the quality of the observation (Fig. 2.14).

All of these above details should be included in your rough observations and your final neat observations. For planets, it is also worth recording the longitude of the central meridian (note some planets such as the gas giants have more than one

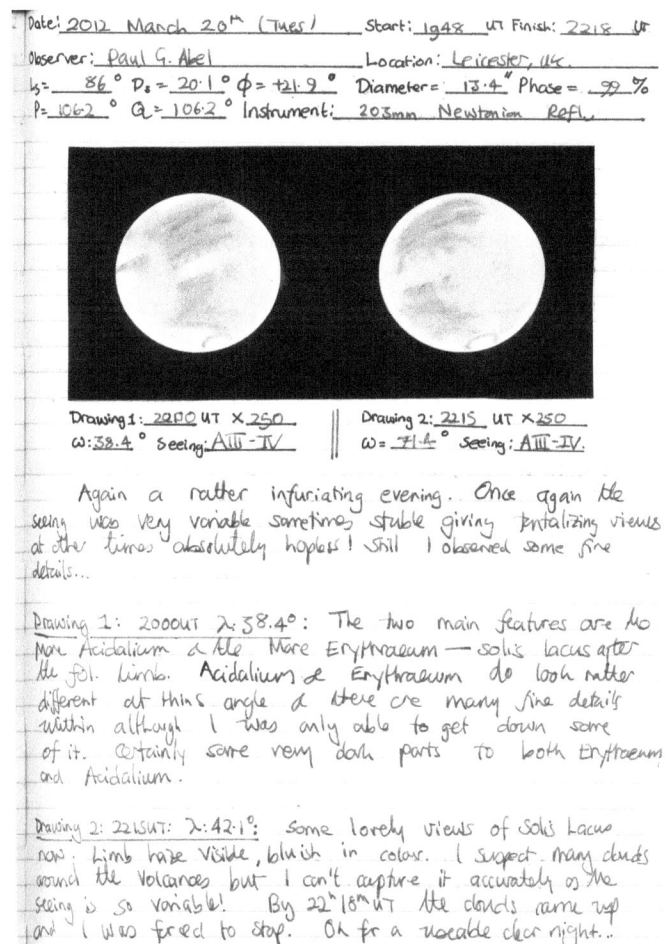

Date: 2012 March 20ᵗʰ (Tues)   Start: 1948 UT Finish: 2218 UT
Observer: Paul G. Abel   Location: Leicester, UK.
Ls=   86 ° Dₑ = 20·1 ° φ = +21·9 ° Diameter = 13·4″ Phase = 99 %
P= 1062 ° Q = 106·2 ° Instrument: 203mm Newtonian Refl.

Drawing 1: 2200 UT ×250   Drawing 2: 2215 UT ×250
ω: 38·4° Seeing: AIII-IV   ω = 71·4° Seeing: AIII-IV.

Again a rather infuriating evening. Once again the
seeing was very variable sometimes stable giving tantalizing views
at other times absolutely hopeless! Still I observed some fine
details...

Drawing 1: 2000UT λ: 38.4°: The two main features are Mo
More Acidalium & the Mare Erythraeum — Solis Lacus after
the fol. limb. Acidalium & Erythraeum do look rather
different at this angle & there are many fine details
within although I was only able to get down some
of it. Certainly some very dark parts to both Erythaeum
and Acidalium.

Drawing 2: 2215UT: λ: 42·1°: Some lovely views of Solis Lacus
now; limb haze visible, bluish in colour. I suspect many clouds
around the volcanoes but I can't capture it accurately as the
seeing is so variable! By 22ʰ18ᵐ UT the clouds came up
and I was forced to stop. Oh for a useable clear night...

Fig. 2.14 A page from the author's Mars log book. All the observing details are recorded, and the neat color drawing is fixed onto the page

of these, as we shall see) so you can keep track of certain features on the surfaces and atmospheres of planets.

One last thing to mention is telescope cooling. If you store your telescope in the house, then it will need time to cool down. When you use your telescope it should be at the same ambient temperature as the environment you are observing in. If it is not, you will find that the air inside the telescope is warmer, and this will create unsteady seeing within the telescope tube (unless it is an open tube system). The telescope will be practically useless, so make sure you give it plenty of time to cool down before you start observing.

Draftsmanship and Artistic Techniques

It is appropriate to mention a few things about drawing in this chapter since it is one of the main activities of the visual observer. Before we start, the first thing to know is that there is a great deal of difference between draftsmanship and art. When producing a work of art, you are free to be as subjective as you like, emphasizing one particular feature and choosing to represent certain features in whatever style you wish. A scientific drawing is the complete opposite to this. It must be made with as much objectivity as possible. There is no room in a scientific drawing (such a drawing of a planet, for example) for the subjective emphasis of certain features. We must put all preconceptions aside and record as objectively as possible what is seen in the eyepiece. We may borrow artistic techniques of shading and coloring, but strictly for scientific purposes.

You do not need any artistic ability to make a drawing (any more than you do to construct a complicated circuit diagram), but you may find some of the drawing techniques used by artists helpful. It might be worth doing a little reading if you feel you are a complete beginner, but to be honest, the only way to hone your draftsman skills is to start drawing at the telescope. The more you draw at the telescope, the better you will become at it. Experience counts for everything. However experienced you become, make sure you never lose your objectivity and draftsmanship. The road from scientific drawing to space artist is a quick and easy one. It is recommended that you learn to draw the planets in black and white.

Do not be disheartened if it takes you a good few months or so to produce a drawing you are happy with. Don't forget, it is not only your drawing skills you are developing but also learning to use your visual system in a more precise way. It does take practice to get the latitudes of the belts right on the gas giants, or accurately place the elusive clouds of Venus, but as you gain more experience and confidence, you will find that your drawing skills will improve quite rapidly.

A final word on the subject of equipment: the user of the equipment is more important that the equipment itself! In fact, for some people, telescopes and flashy equipment are of more interest than actual observing The important thing is that you find a telescope you can use when you want to and you come to understand its abilities and limitations. Don't be put off by the goading of other people to obtain some enormous light bucket. It's easy to fall into the trap of buying 'bigger and better' (aperture fever, as it has come to be known), but when that happens you will spend most of your time learning to use a new telescope rather than sharpening your observing skills.

Chapter 3

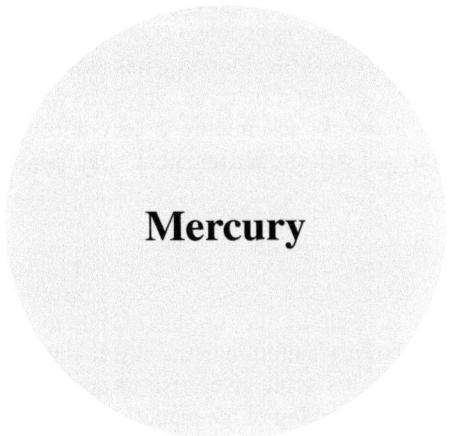

Mercury

Distance from the Sun: 69.7 million km (max), 45.9 million km (min)
Diameter: 4,878 km
Sidereal Period: 87.969 days
Rotational Period: 58.6461 days
Mean Surface Temperature: 350 °C (day) −170 °C (night)
Apparent Diameter: 12.9″ (max), 4.5″ (min)
Albedo: 0.06
Maximum Apparent Magnitude: −1.9
Satellites: None

A Brief Look at Mercury

Mercury gets its name from Roman mythology. Mercury was the messenger of the gods. This planet was one of the five bright planets known since antiquity, and of all the planets in our Solar System, it is Mercury that possess the greatest challenge to amateur astronomers.

There are two problems to be overcome. First, Mercury is the closest planet to the Sun, and as a result, it never strays far from the Sun in the sky. It appears either as a morning or an evening star and is always very low down. It can never be viewed in a dark sky, and when at its best it will rise (or set) about an hour and a half before (or after) the Sun.

P.G. Abel, *Visual Lunar and Planetary Astronomy*, The Patrick Moore
Practical Astronomy Series, DOI 10.1007/978-1-4614-7019-9_3,
© Springer Science+Business Media New York 2013

Secondly, the planet is only a little larger than the Moon, and as a result, it never exceeds 13″ in size! This, combined with its rapid motion about the Sun, means that it is only really a good naked eye object for a couple of weeks during an elongation the period during which the planet is visible.

At first glance, it seems that little Mercury has little going for it in terms of amateur observation. However you'd be quite wrong to write it off just yet. As we shall see, the planet is a deeply fascinating object with all sorts of wonderful mysteries, and a skilled and determined amateur may contribute a great deal in solving these mysteries.

Orbit and Elongations

In order to get a feel for Mercury's motion in the sky, it is worth examining it's orbit a little closer- especially as it's orbit gives rise to some interesting behavior. In Fig. 3.1, I have given a rough drawing of the planet's orbit. Essentially we can break Mercury's appearance in the sky into two periods- a Western Elongation, when the planet is visible in the morning sky and Eastern Elongation when the planet is visible in the evening sky. The two elongations are separated by conjunction – inferior conjunction when the planet passes between the Earth and the Sun, and superior conjunction when then planet is behind the sun.

Starting at position 1, Mercury is behind the Sun and is said to be in superior conjunction. The planet is not visible from Earth during this time. The planet then moves in it's orbit towards position 2 to start an eastern elongation. During this time, the planet appears to viewers on Earth as a waning gibbous and is visible in the evening sky. As can be seen in the diagram, the planet is coming closer to the Earth, and so it's apparent size increases.

The planet continues in it's orbit until it reaches Position 3, the point of 'dichotomy' At this point it appears to be half illuminated- just like a half moon. The planet then moves to position 4 where it appears to have a waining crescent. At around this point we arrive at greatest eastern elongation and the planet is at it's furthest point from the Sun as seen in the skies of Earth. The disk size is still increasing.

Eventually the planet moves into position 5 where it is effectively between the Sun and the earth- this is the point of inferior conjunction. Usually, Mercury passes just above or below the Sun, but there are rare occasions when the planet actually passes in front of the Sun in a phenomena known as a transit. During this time it can be viewed as a small black disk crossing the face of the Sun.

After passing through inferior conjunction, the planet moves into the morning sky and embarks on a western elongation an. At position 6 it is now has a waxing crescent and near here, greatest western elongation is reached and once more Mercury (from our point of view) at its furthest point from the sun. The planet is now moving away from us and so it's apparent size and magnitude decreases and at Position 7 it is again at dichotomy. After this point, Mercury will continue to

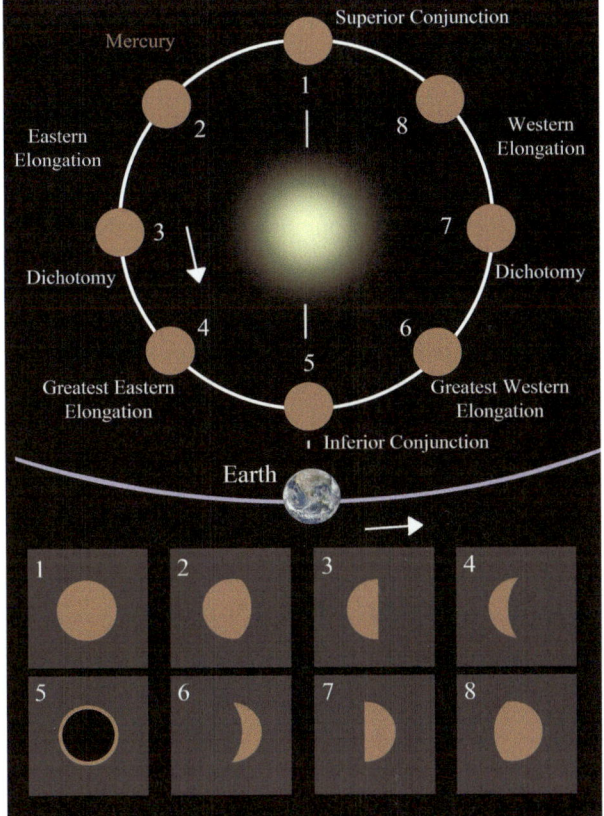

Fig. 3.1 The orbit of Mercury along with a view of the planet as seen in a telescope (with north at the *bottom*)

decrease in size (as viewed from Earth), as it reaches position 8, and now presents a small waxing gibbous disk.

Finally the planet passes behind the Sun and reaches position 1. The planet is at inferior conjunction, and embarks on a new western elongation and the whole cycle begins again. I should mention in fact that not all elongations are equally favorable.

From the perspective of the Northern hemisphere favorable Eastern elongations occur in spring, while western ones occur in the autumn. As we shall see, Mercury is best observed during day light time, and this the time when most amateurs make their observations of the planet.

If we are to make useful meaningful observations of Mercury, it is first best to understand a little bit of how the planet is made. Obviously there isn't the space here to give a full examination of our current understanding.

Mercury as a World

If we are to make useful meaningful observations of Mercury, it is first best to understand a little bit of how the planet is made up. Obviously there isn't the space here to give a full examination of our current understanding, but a short review of the pertinent facts will be most helpful.

Broadly speaking, the planets in our Solar System fall into three categories: terrestrial planets, gas giants, and ice giants. Mercury is a terrestrial planet, a small rocky planet with a mantle and core. (Gas giants such as Jupiter and Saturn are predominantly hydrogen, while the ice giants such as Uranus and Neptune are largely made of Hydrogen and Helium frozen methane and ammonia).

Mercury is a rather dense sort of planet, and given its size, we can infer from this the fact that the planet must contain a large metallic core. *Mariner 10*, in the summer of 1971, became the first spacecraft to reach the planet and show us the surface in any detail. It became obvious that Mercury's surface strongly resembled that of our own Moon – large impact craters, planes and escarpments were to be found all over the planet.

Since *Mariner 10*, the MESSENGER of spacecraft have visited the planet and filled in some more details for us. As expected, there was a small, almost negligible atmosphere, the main components of which are sodium, oxygen helium and potassium. These exist in a thin gaseous state, and it is unlikely that such a tenuous atmosphere could produce anything as dramatic as clouds.

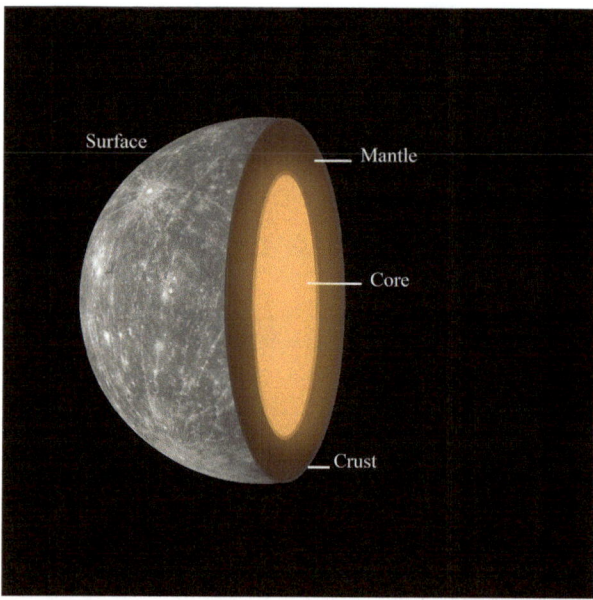

Fig. 3.2 The interior of Mercury

Figure 3.2 is a cutaway diagram showing the interior of Mercury. The heavily cratered surface is just the top of the solid rocky crust, which is made largely of silicates. It is thought that the crust is somewhere between 100 and 300 km thick. Beneath this we have a molten layer of rock and iron called the mantle, and finally we have the large iron-rich core, thought to be some 1,800 km in radius.

It is clear that Mercury's core is very large indeed, certainly the largest of any of the terrestrial planets, and how it came to posses such a large iron-rich core is still a mystery. The current favored theory is that in the remote past, when the Solar System was still being formed, the planet collided with a planetesimal – a small Moon-sized object – and the resulting collision tore away much of the surface, leaving a much reduced crust with the large core inside.

When we look at Mercury through a telescope, we are looking down on to the cratered surface, and as we shall shortly see, early telescopic explorers of the planet were able to observe albedo features. Without a doubt, the most dramatic feature on the planet is the Caloris Basin: an enormous impact crater some 1,550 km in diameter. It is thought that some large object collided with the planet in the past, and the resulting impact was so large, it sent shock waves around the planet that are frozen into the stone on the other side of Mercury.

Until recent times, only *Mariner 10* had visited Mercury in 1975, and virtually all of our information came from this glancing visit. Although the other planets of the Solar System have enjoyed visits from a impressive range of spacecraft, Mercury has languished in the blinding glare of the Sun, relatively obscure. There is good reason for this. The Sun's strong gravity that close to Mercury makes it harder to get there, and with the Sun being so close, the environment is very hostile. The neglect has been put right with the arrival of MESSENGER.

The MESSENGER spacecraft (which stands for Mercury Surface, Space Environment, Geochemistry, and Ranging Spacecraft) was launched in August 2004 and took a somewhat convoluted route to Mercury, passing Earth once, Venus twice, and Mercury itself three times before finally settling into orbit on 2011, March 18.

MESSENGER has made a number of startling discoveries about the planet, perhaps the most remarkable of these being the existence of water ice and organic compounds on the north pole of the planet. Moreover the planet has a substantial magnetic field. This field is strong enough to deflect the constant stream of energetic particles emitted from the Sun (the solar wind) and create a magnetosphere. It has provided further evidence that the planet has a substantial core, probably comprising about 85 % of the planet. Moreover, it has mapped the entire surface and produced a comprehensive 3D map of Mercury's surface. NASA has extended the mission for another year so MESSSENGER will continue until at least March 2013. One thing is certain, the spacecraft has demonstrated that Mercury is far from being a boring dead rock in space.

Figure 3.3a shows the heavily cratered surface of the planet in wonderful detail, and Fig. 3.3b shows stunning details of the bright rays of Mena Crater.

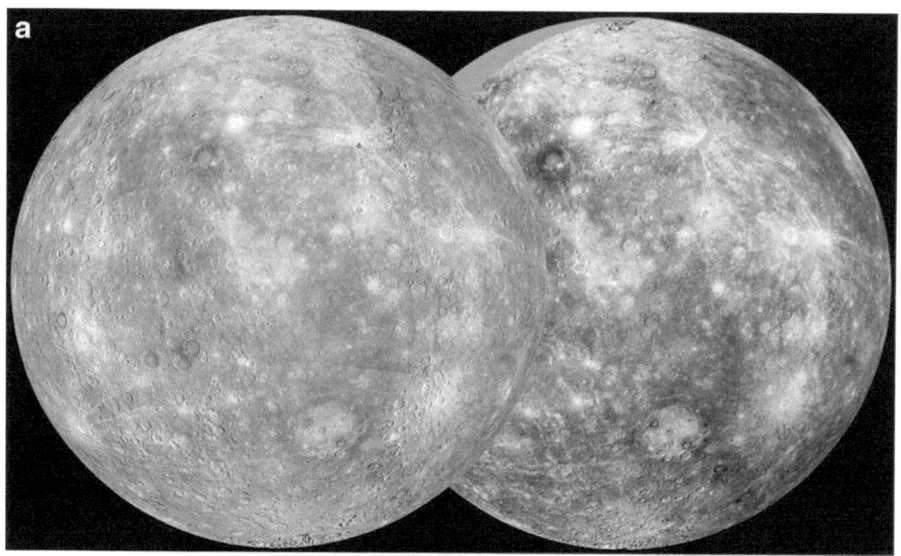

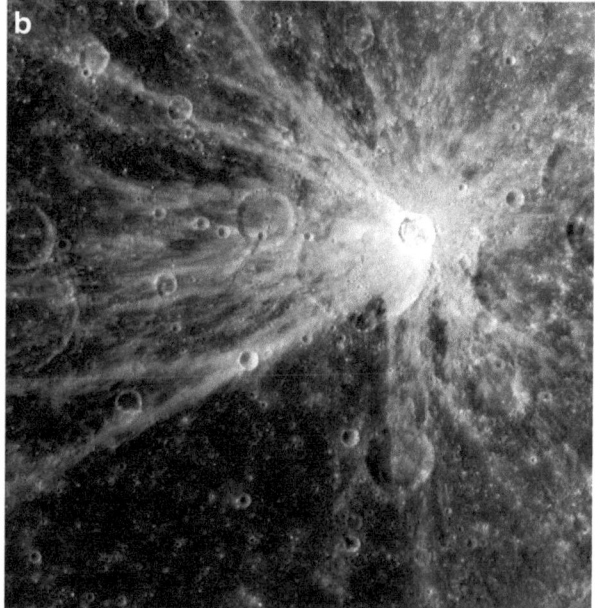

Fig. 3.3 (**a, b**) Two MESSENGER images (Courtesy of NASA)

Although MESSENGER has provided all of this wonderful science about Mercury, we already had a limited knowledge of the planet thanks to the work of early telescopic observers, and it is their observations that we shall now examine, for if we are to make useful observations in the future, it is clear we must see what observers have done in the past.

Historical Observations

When at its brightest, Mercury is a conspicuous object low down in the sky, and so it was not too long after the invention of the telescope that the early Solar System explorers pointed their telescopes at the planet. Much of what they saw then is identical to what we see today, but some of it was not.

A number of the early observers saw bright 'stars' and glows on the surface of the planet, and many of the more interesting observations of the past can no doubt be put down to poor optics, for example Wurzelbaur's observation of a gray spot on the planet on 1697, November 3, or De Plaatde's observation of a bright luminous ring around the planet on 1736, November 11, seem to be the result of poorly aligned optics.

Having said that, it is the case that some phenomena like such as the "blunting of Mercury's horns" – an interesting effect where the planet's extreme north and south points (in crescent stage) appear rounded rather than sharp points – were observed reliably by astronomers as recently as the 1950s. So let us review some of these puzzling observations from the past.

Let us first deal with the great telescopic observer Johann Hieronymus Shröter (1745–1816). Shröter was a German astronomer, who, after hearing of Hershel's discovery of the planet Uranus, decided to abandon his job in legal practice and devote himself entirely to the study of astronomy.

Shröter moved to the small town of Lilenthal, where he became chief magistrate and governor of the region. He established a modest observatory and used a variety of telescopes, the largest being a 16.5 cm refractor, which he used to chart the lunar surface and make extensive drawings and observations of the planets. Shröter made many important contributions to astronomy and soon established himself as a reliable observer, but alas his examination of the Solar System was not to last. In 1813 the Napoleonic wars, which had been raging in Europe, arrived at his door. The brutal General Vandomme destroyed Shröter's observatory and his books and records of observations. Poor Shröter, never recovered from the disaster.

Long before then, however, on the 1799, May 7, Shröter and his assistant Harding were observing Mercury when they noticed two spots on the surface and a sort of luminous halo around the planet. A few months later, on 1800, March 26, both Shröter and Harding observed that the horns or the north and south points of Mercury were not identical. Rather the southern part of the crescent appeared to be decidedly blunt, while the northern one appeared as a sharp point.

Shröter was to make one more puzzling observation of Mercury; on 1814, March 17, he observed what he believed to be a bright point of light on Mercury's south pole. He had of course seen a near identical effect on the Moon – the result of a mountain peak catching the sunlight while the rest of the mountain remained in shadow. Very soon he realized this could not be the case, for such a mountain would need to be over 18 km high! He realized this was most unlikely given that the planet was smaller than Earth!

Observations of blunted horns and bright stars continued to be made however; on 1802 November 9, the astronomer Fitsch and others observed a white spot on

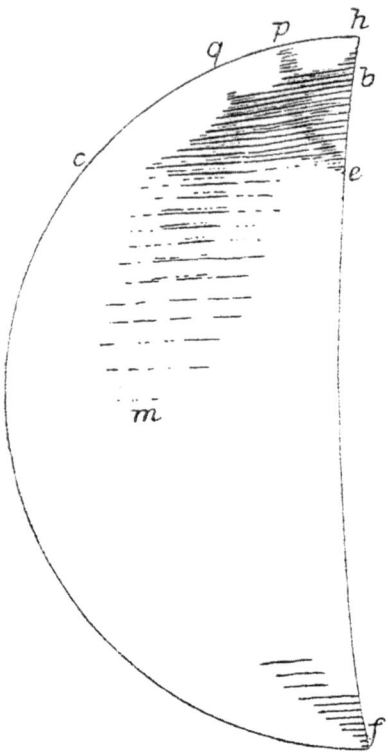

Fig. 3.4 Schiaparelli's drawing of Mercury on 1895, July 13, at the Brera Observatory. Some *dark* markings are indicated (Published in Schiaparelli's *Astronomische Nachrichten* Bd 138. No 3304 (1895))

the disk of Mercury, and Humboldt noted during a solar transit that the disk of Mercury was perfectly sharp, indicating that there was no substantial atmosphere around the planet, for had there been such an atmospheric envelope the planet's limb would have been fuzzy, not sharp and well defined.

After Shröter, the next serious astronomer to pay the planet any attention was Giovanni Schiaparelli. Schiaparelli spent most of his professional career at the Berera observatory in Milan. He would later become famous for suggesting that the Martian surface was covered in a series of geometric lines and shapes he called "canili" – an idea Percival Lowell took up with wanton abandon (Fig. 3.4).

Schiaparelli used both a 8.6 in. and 11.4 in (220 & 290 mm) refractors at the observatory and tended to observe during the day when the planet was higher in the sky and not so troubled by poor seeing. Schiaparelli was frequently able to see fine albedo markings, and over time, he made some 150 drawings of the planet. He later compiled these drawings to produce a map of the planet; indeed Schiaparelli's map, and the one compiled by the legendary observer Antoniadi remained the two main reference maps of the planet until the advent of spacecraft.

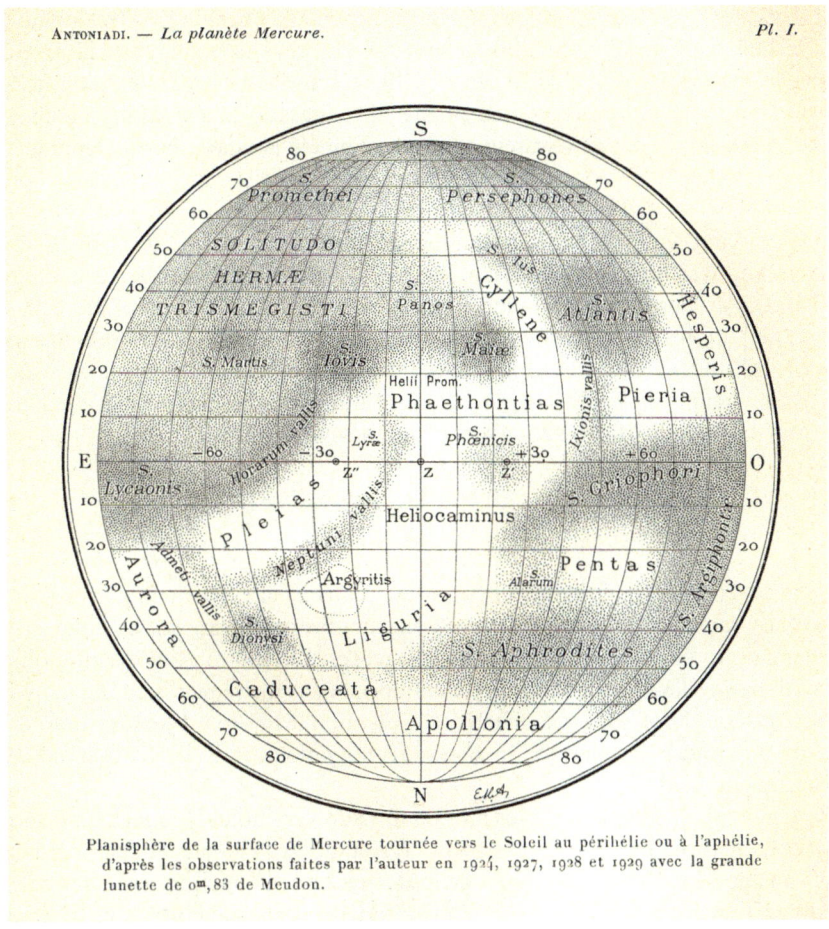

Antoniadi. — *La planète Mercure*.   Pl. I.

Planisphère de la surface de Mercure tournée vers le Soleil au périhélie ou à l'aphélie, d'après les observations faites par l'auteur en 1924, 1927, 1928 et 1929 avec la grande lunette de 0ᵐ,83 de Meudon.

Fig. 3.5 A map of Mercury drawn by the legendary observer E. Antoniadi based on observations made of the planet in 1924, 1927, 1928 and 1929 (Courtesy E. Antoniadi)

By the end of the 1800s, the celebrated author and observer William Frederick Denning had begun observing the planet. Denning was a dedicated amateur who used his 10 in. (250 mm) reflector to observe the Moon and planets at his home in Bristol, England. He was one of the first people to write a popular astronomy book aimed at serious amateur astronomers, and his book *Telescopic Work for Star Light Evenings* is an invaluable reference even today.

It was on the evenings of 1882, November 5, 6, 8 and 9, that Denning made some fascinating observations of Mercury, a number of which were confirmed by Schiaparelli in Milan. Denning sent Schiaparelli his observations. On November 5, using his 10 in. (250 mm) reflector at 212×, Denning was able to observe a series of dark spots and a well marked luminous area between the E.N.E part of the limb and the terminator. He also noted that the southern horn was distinctly blunted.

He repeated his observations on the November 6 or 8, where in good seeing he was able to once again make out the bright spots of the previous night, but within one of the bright regions he was able to see a brilliant star-like point. In addition he also observed on the SW border of one of the light areas a dark marking along with another, darker in tone, that was inlaying the blunted southern horn. Denning was again able to repeat his observations on November 9 and see much the same sort of thing, only now the features had moved slightly eastwards, suggesting they were real features moving with the planet's rotation. In more recent times, Richard Baum of Chester, who was for many years the director of the Mercury and Venus Section of the British Astronomical Association, made a number of observations of Mercury and is fairly certain that some of the features on Mercury do vary a little with intensity.

Although there is no way of knowing how many of these observations were of real phenomena, it is worth keeping them in mind, for the answers of puzzles of yesterday often lie in the observations of tomorrow.

Observing Mercury

You can observe Mercury in one of two ways. You can catch it when it is prominent in the early morning or evening sky, or you can observe during the daytime. If you are planning to observe Mercury when it a reasonable morning or evening object, you will need to find yourself a fairly clear horizon, as the planet is appreciably low down. In the case of spring elongations, you should observe Mercury before the time of maximum elongation. In autumn elongations it is best observed after maximum elongation.

Remember that Mercury moves around the Sun very quickly indeed, and so you will notice that the planet changes phase very quickly. The planet can be a bright object in the sky 10 days before maximum eastern elongation, then practically invisible a few days afterwards! Remember, too, the associated problems of observing a object when it is low down. The light from an object is having to pass through more of Earth's turbulent atmosphere, and so the seeing quality is liable to be rather poor. Views of the planet in such conditions are bound to disappoint; indeed all that may be seen is a small blank disk!

Alas, then, only real way to observe the planet is during the daytime, when the planet passes higher through the sky, but to do this you will need a telescope equipped with setting circles, i.e., an equatorial mounted telescope equipped with right ascension and declination. Always remember that the planet will be close to the Sun, and so here's a word of warning: NEVER point your telescope in the vague direction of the Sun in the chance of spotting Mercury. If you should catch the Sun, the intense heat and light will blind you. Even a moment's look through a viewfinder is enough to do a lifetime's worth of damage. The only safe way to observe Mercury in the daytime is to use the setting circles (see Chap. 2). If you have a program such as *WINJUPOS*, all you need do is find the R. A. and dec. of Mercury then dial this up on your telescope. This is really the best way of observing the planet.

Remember that Mercury is a small world, only a little larger than the Moon, so it never subtends a diameter larger than 12.9". Powers between 180×–250× are really best suited for observing Mercury. Initially you will probably see a small orangish-colored disk, but as you spend time looking at the planet, you will see there are darker sections and lighter regions all over the disk.

By observing over a whole elongation, you can observe all of Mercury's disk. Although it keeps the same face towards the Sun, it does not keep the same face turned towards us, so as it moves around the Sun, you gradually get to see the whole of the surface.

With this in mind, let us now turn our attention to some prominent surface features.

Surface Features

If you wish to make serious long-term visual studies of Mercury then you will need to become familiar with the albedo features that can be seen from Earth with anything from a modest-sized telescope upwards. To help with this, look at Fig. 3.5, a map drawn up by the visual observer Mario Frassati using his 203 mm (8-in.) SCT telescope, and reproduced here with his permission (Fig. 3.6).

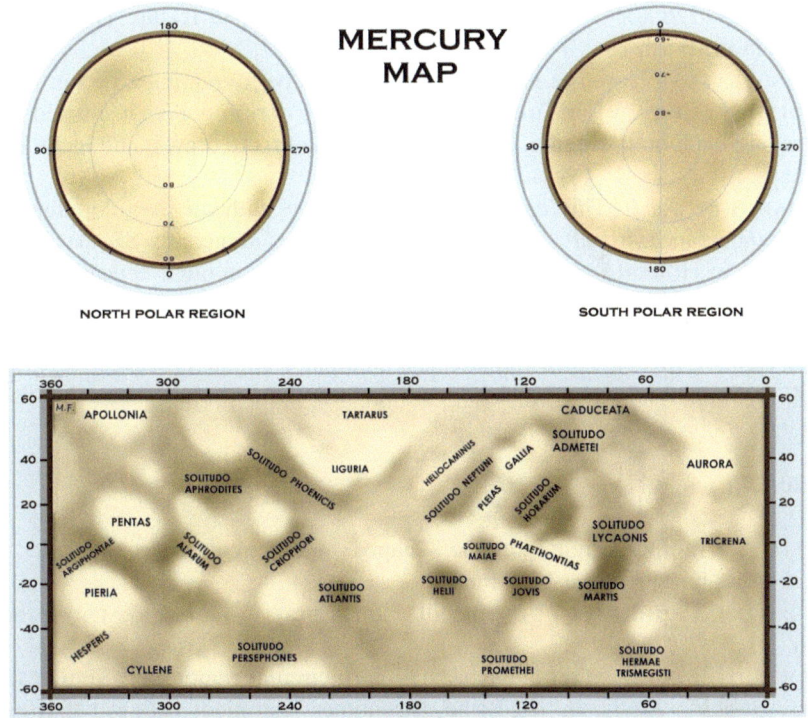

Map of regolith albedo features of Mercury based on 144 visual observations by MARIO FRASSATI at Crescentino (VC), Piedmont, ITALY, between 1997 January and 2006 January. Instr. :203 mm f/10 SCT x250–400. *Mario Frassati*. (Annotated with IAU nomenclature)

Fig. 3.6 Map of Mercury made by Mario Frassiti using his 203 mm SCT ×250–400. Note that north is at the *top* (Courtesy Mario Frassiti)

Let us now review some of the prominent features. We shall do this by looking at features in three blocks of longitude λ: 0–120°, λ: 120–240° and λ: 240–360° (the Greek letter lambda 'λ' is used as an abbreviation of longitude). Obviously, the features that are visible depends on whether we are looking at Mercury in an eastern elongation (evenings) or a western elongation (mornings). Included is the letter 'E' if a feature is best viewed at an eastern elongation, or a 'W' for a western one.

λ: 0–120°

SOLITUDO HERMAE TRISMEGISTI (W)This is a small dark region very near the south pole. It can be seen with a 6 in. (150 mm) reflector and looks like a dark curved feature.

TRICRENA (W)A small dark grayish circular feature situated on the equator.

AURORA (W)This is a reasonably bright region located in the north. On first glance it appears circular, but on further inspection a lighter diagonal section appears to join it in the north.

SOLITUDO MARTIS (W)A reasonably dark elongated feature just below the equator in the south.

SOLITUDO LYCAONIS (W)A fairly dark irregular shaped feature just above the equator in the north.

CADUCEATA (W)A rather vague lighter region to be found in the far north

SOLITUDO JOVIS (W)This usually appears to be a dark round spot situated below the bright Phaethontias region. Schiaparelli was the first to draw it and include it on his maps.

PHAETHONTIAS (E)This can be a bright feature with an irregular dumbbell-shaped appearance. It starts on the equator and moves diagonally SE.

SOLITUDO HORARUM (W)This is a dark region that can often appear as a broad dark streak. It is a fairly conspicuous object indeed a number of people have claimed to seen it with a small instrument.

SOLITUDO ADMETEI (W)A rather vague dark patch situated between the brighter Gallia region to the east and the Caduceata region in the north.

λ: 120–240°

GALLIA (W)Another bright dumbbell-shaped region in the north, smaller than Phaethontias but of similar brightness.

SOLITUDO PROMETHI (W)Another ill defined grayish region located in the far south.

SOLITUDO MAIAE (E)Yet another small dark circular region very similar to the nearby Solitudo Jarvis. In 1927, the astronomer Antoniadi believed the feature became temporarily obscured.

PLEIAS (W)A fairly bright region located between the Solis Neptuni and Solitudo Horarum.

SOLITUDO NEPTUNI (W)This has the appearance of a long dark strip starting below Gallia and ending just above the equator in the north.

HELIOCAMINUS (E)Another bright elongated region in the north.

SOLITUDO HELII (E)Another small dark circular patch like Maiae and Jovis.

SOLITUDO PHOENICIS (E)A broad grayish region to be found in the north under the Liguria region.

LIGURIA (W)A rather vague, ill-defined light grayish region in the north. A brighter patch has often been observed in the center of this region.

λ: 240–360°

SOLITUDO PERSEPHONES (E)This is a rather extensive feature, and under good conditions some fine details can be seen here. It is a very dark feature, and Schiaparelli believed it to be the cause of the occasional 'blunting' of the southern horn when the terminator was in this region. If this is the case repeated observation should confirm this.

SOLITUDE CRIOPHORI (E)This is a very dark feature and widely considered to be the most prominent feature visible on the disk.

SOLITUDO APHRODITES (E)Another prominent dark feature, it appears to be darkest along its southern boundary.

SOLITUDO ALARUM (E)A reasonably dark feature situated to the east of the bright Pentas region.

PENTAS (E)A fairly bright oval-shaped region just above the equator in the northern hemisphere.

SOLITUDO AGRIPHONTAE (E)A dark narrow region between the brighter Pentas and Pieria regions. It lies on the equator, and a number of people claim to have seen it with small instruments.

PIERIA (E)A bright oval region in the southern hemisphere. It is located between the darker Solitudo Argriphontae on the equator and the lighter Hesperis region in the south

APOLLONIA (E)A vague peanut-shaped region, fairly bright, located in the far north.

HESPERIS (E)A rather vague dark feature just to the south of the bright Pieria region.

CYLLENE (E)A vague circular dark patch situated in the far south. It is connected to the darker Hesperis region.

Drawing Mercury

Unlike some of the other planets in the Solar System, Mercury is relatively easy to draw and record. It is worth doing some preparatory work before going outside. For example you can use *WINJUPOS* to determine what the current phase of the planet is, and by finding the CM longitude (i.e., what line of longitude is on the central meridian when you are observing) you can work out from the map what you can expect to see. To draw Mercury, you should use a disk of 50 mm diameter. Some people prefer do draw the phase in for themselves, but included in the appendix of this book is a set of blanks which have the phases for Mercury and Venus. Simply use the blank that matches the phase on the day you are observing.

Once you get outside, the first thing to do is to spend some time working out what magnification is suitable – and this will depend upon your seeing conditions. If the seeing is good to average, consider using a power of 210× or more. When you have done this, spend about 10 min observing what is on the disk. When you are reasonably certain of what you can see, start making the drawing.

When making a drawing, it is best to do it in a methodical fashion: start by putting in any shading or markings on the terminator and the limb, then work from south to north putting in the fine details. If you should see anything worth remarking – a bright feature, say, or a slight change in appearance of a feature, make a note of it. When you are done, record the date, time (in UT), and seeing conditions, and don't forget to record the instrument and magnification as well. This will all be extremely useful information later when you come to make a neat copy of your observation. You might also record the phase and CM longitude for Mercury as well, as this can be important when working out what features were seen later on.

It is worth noting that it is best to turn your rough observations into neat drawings as soon as possible, for the longer you leave it, the more the errors are likely to creep in!

Advanced Projects

If you wish to make more detailed studies of the planet, then you can try some of the following projects.

Map Making

A large collection of disk drawings are very useful, as you can summarize your observations of the planet's surface by making a map. In the case of Mercury, you need to collect drawings from both eastern and western elongations if you are to map the entire surface.

The first thing to do is draw out a blank rectangle with 0–360° marked on it in equal longitude and with latitude ranging from −60° in the south and +60° in the north (with 0° being the equator and center of your map). Then, go through your drawings, and try to pick out some regular features and set intervals of longitude. For example find a prominent object at 0° longitude, then at 60°, then at 120° and so on. When you have done this, simply draw them onto the map at the correct latitude and longitude. This will take a bit of practice, and normally the temptation is to make the objects too large!

When you have the prominent features in, you can use these as base points to put in the finer details you have observed, and before long, you'll have a good visual summary of your observations. The final touch is to add the names of the features you have observed. In the case of planets where dynamic changes occur (such as in the case of Mars), it is nice to look back at your maps over the years to see if any features have changed in a subtle way (Fig. 3.6).

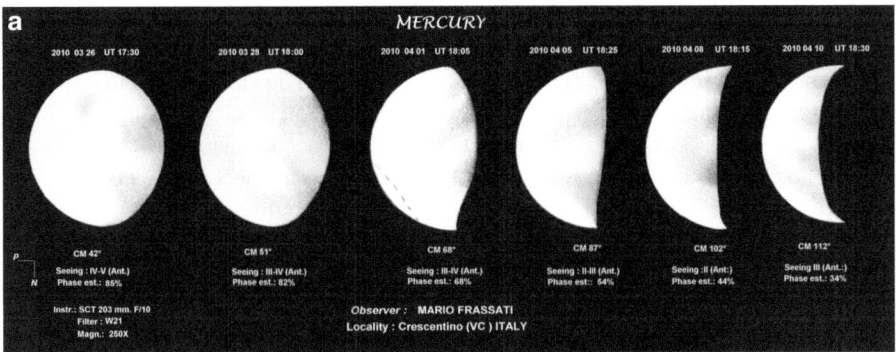

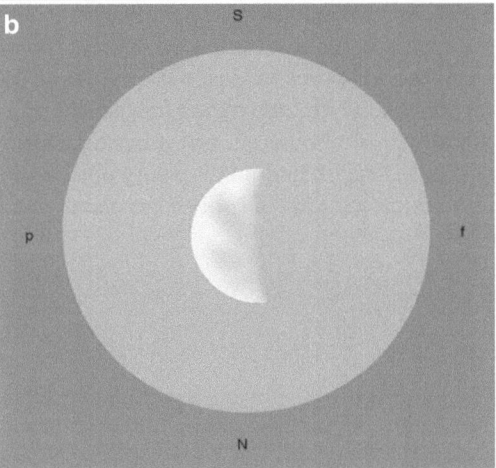

Fig. 3.7 (**a**) An excellent series of drawings by Italian observer Mario Frassati during March and April of 2010. Frassati used a 203 mm SCT ×250 and a orange W#21 filter (other observing details indicated on the drawing). (**b**) A challenging daytime observation by U. S. observer Carlos Hernandez. The observation was made on 2009, August 22, at 2000UT with a 203 mm Maksutov Cassegrain at a power of ×286 (Courtesy Mario Frassiti and Carlos Hernandez)

Intensity Estimates

These are a useful way of keeping track of subtle changes in the brightness of a feature. The scale adopted by the British Astronomical Association is as follows: 0 = extremely bright spots, 1 = bright regions, 2 = general hue of disk, 3 = some darker shading near the limit of visibility, 4 = dark shadings, and finally 5 = extremely dark shadings.

When making an intensity estimate you can either make another disk drawing and assign each of the features on the disk a value between 0 and 5, or you can make a table in your notebook and list the feature and put their intensities by the side. In either case, don't forget to record the usual date, time and instrument used in the case of all the planets except Venus, where you might to prefer to use a table simply because you can transfer the numbers at the end of an elongation into a spreadsheet and work out the mean and standard deviation of intensity estimates of the features observed. After a number of years, this becomes a reliable way of tracking any subtle changes in brightness/darkness of the surface features of planets.

Albedo Changes

Although it seems to be the case that Mercury is a fairly inert sort of world, nevertheless, it is certainly worth trying to repeat observations of the past (for example, the blunting of Mercury's horns) and checking back to see if there have been any long-term changes to any of the features seen on the disk, no matter how subtle.

Mercury is a difficult world to observe, and it has long since been neglected by amateurs who label it as a difficult and unrewarding object. This, as we have seen, is quite untrue, and who knows what surprises the planet has waiting in store for the dedicated amateur.

Chapter 4

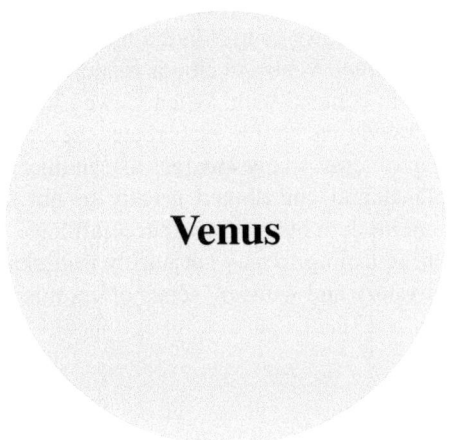

Venus

Useful Facts

Distance from the Sun: 109 million km (max.), 107.4 million km (min.)
Diameter: 12,104 km
Sidereal period: 224.701 days
Synodic Period: 583.9 days
Rotational Period: 243.018 days
Mean Surface Temperature: 467°C
Apparent Diameter: 65.2″ (max.), 9.5″ (min.)
Albedo: 0.76
Brightest Apparent Magnitude: −4.4
Satellites: None

A Brief Look at Venus

When seen in the early morning or evening sky, Venus has an elegant beauty unmatched by any other naked-eye celestial object, shining as it does with a wonderful silvery effulgence. When seen in a twilight sky it looks like a bright lamp floating in a lake.

It would be easy to think that given its relative closeness and apparent brightness the planet would be an easy target to observe, yielding quite readily to telescopic scrutiny. Alas, this is not the case, and out of all of the planets in the Solar System, Venus is probably the most difficult of the lot. This does not mean it is impossible

P.G. Abel, *Visual Lunar and Planetary Astronomy*, The Patrick Moore
Practical Astronomy Series, DOI 10.1007/978-1-4614-7019-9_4,
© Springer Science+Business Media New York 2013

or indeed not worth bothering with. Rather it means that we must take a different approach to the planet.

So what is it that makes Venus such a challenging object to study? The answer is, of course, the very thing that draws us to it in the first place – its brilliance and the cause of it, the clouds. The thick Venusian clouds reflect a lot of light, which means that there is a lot of glare to contend with. When viewed in the morning or evening sky, the planet is rather low down, so this exacerbates the situation further.

In spite of all these problems, many amateur astronomers have made valuable observations of the planet and contributed greatly to our understanding of the planet. It is only in recent modern history that spacecraft have visited the planet and built on this foundation, revealing to us what a truly remarkable world this planet is, and if you're after mystery and wonder, Venus offers this in great abundance.

Orbit and Elongations

Venus is the second planet out from the sun (in order of distance), and so like the planet Mercury is classed as an inferior planet. Venus moves in a practically circular orbit around the Sun, taking 224.7 days to complete one orbit. As the planet lies between the Earth and the Sun, this means that, like Mercury, Venus has a phase cycle.

Figure 4.1, shows the path of Venus around the Sun, along with the appearance of Venus as it appears through a telescope at various different positions. Starting then at position 1, Venus is said to be at Superior conjunction, it is behind the Sun and as viewed from Earth, it appears as a fully illuminated disk very close to the Sun in the sky.

The planet then moves around in it's orbit, and is visible in the evening sky and starts on eastern elongation. Slowly, the planet's apparent size increases as Venus draws closer to us, it reaches position 2. In the telescope, it now appears distinctly gibbous, and is noticeably larger in size than when it was at position 1. Venus continues to position 3, at this point the planet is said to be at dichotomy. In the telescope it looks like a half-moon. The planet continues on her path coming ever closer to us, until it reaches position 4. By now, the planet appears as a waning crescent, and the disk size is rather large. At this point Venus is said to be at greatest eastern elongation, and has reached its furthest point from the Sun as seen in our skies. Venus continues to come closer to us it's phase decreasing but it's disk size increasing and it finnaly arrives at position 5 when it said to be at inferior conjunction, and effectively lies between the Earth and Sun.

Inferior conjunction marks the end of the eastern elongation, and as the planet is at it's closest point to Earth, the disk size can be an enormous 50"; the phase however is tiny at this point, with only a tiny amount of the planet illuminated. After inferior conjunction, the western elongation starts and Venus returns to the morning skies. It moves on to position 6 where it appears as a waxing crescent in the telescope and is at greatest western elongation and rises many hours before the Sun. The planet is moving away from us now, and so the apparent disk size is decreasing. It continues

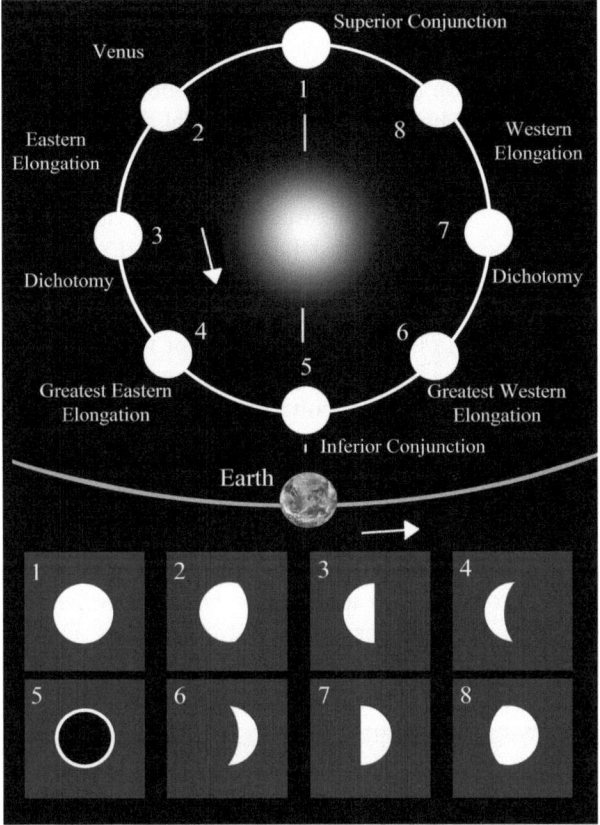

Fig. 4.1 The orbit of Venus around the Sun and its appearance in the telescope (north at the *bottom*) at various positions. Drawing is not to scale

in it's orbit and reaches dichotomy at position 7. Venus continues to move on to position 8 where it now has a waxing gibbous appearance, and continues on back to position 1 where it is at superior conjunction and the whole cycle begins again.

On average, it takes Venus some 583 days to go from Superior conjunction to inferior conjunction, and this period of time is called the Synodic period.

Now, it may be thought that, at the time of Superior conjunction Venus passes behind the Solar disk and is hidden by the Sun, while at times of Inferior conjunction, the planet transits the solar disk. This however is not the case, and this is because the orbit of Venus is inclined by 3 degrees and 4 minutes. As a result of this, the planet more often than not passes above and below the sun at the times of conjunction, and only rarely passes behind the Sun or passes in front on the Sun. So, we have seen how Venus moves through our skies, alternating between Eastern and Western elongations let us now examine some of the interesting facts about the planet which have come to light.

Features of the Planet

It is not without good reason that Venus has been dubbed "the planet of mystery." Like Mercury, Earth and Mars, Venus is a terrestrial planet. However, it has the thickest and densest atmosphere of all the four inner planets, and those thick clouds never allow us to see the surface. Before spacecraft revealed the truth everything was imagined to exist under the thick curtain of cloud – vast swamps and oceans, perhaps with dinosaurs, or maybe vast lakes of oil and other hydrocarbons, or maybe the surface was a continual desert. All these suggestions were given serious attention at some time or another.

Figure 4.2 shows the currently accepted model of the internal structure of the planet. We have at the top the thick atmosphere; below this we have the surface and crust. Under this is believed to be the molten mantle with a hot core at the center of the planet. Interestingly there is no detectable magnetic field.

The atmosphere was one of the first things discovered by telescopic observers; in 1761 the Russian astronomer Mikial Lomonosov observed Venus transiting the Sun. As he watched the disk of Venus pass onto the solar limb, he was surprised to observe a bright ring of light surrounding the planet. The only conclusion could be that the planet was surrounded with a thick atmosphere – as when Mercury transited the disk, the outline of its orb was crisp and sharp, indicating no atmosphere.

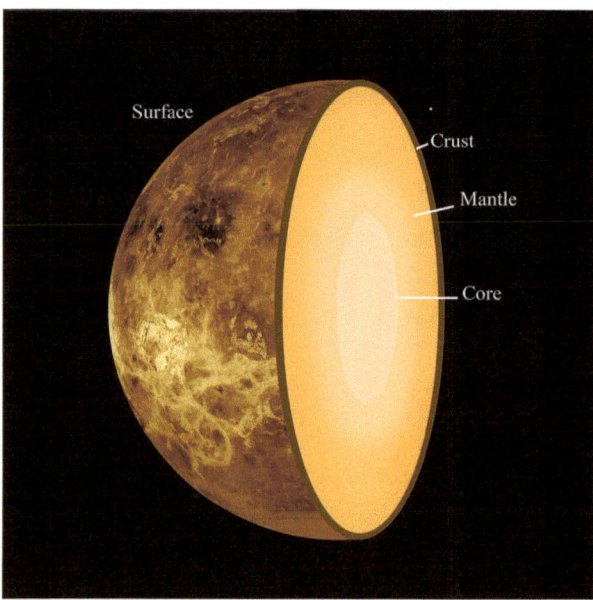

Fig. 4.2 Cutaway diagram showing the currently accepted model of the structure of Venus (cloud deck not shown)

Although we had to wait until the advent of the space age to see the surface of Venus, Earthbound astronomers were able to make some progress with studying the atmosphere. Two astronomers, E. Pettit and S. B. Nicholson, used the 100-in. reflector at Mt. Wilson to study the Venusian clouds. They found that temperature at the cloud tops was around –33 °C (an accurate assessment), and that the main constituent of the atmosphere was the heavy gas carbon dioxide (CO_2). (This of course led people to suppose that any oceans on the planet may be carbonated, like fizzy mineral water!) The presence of vast amounts of carbon dioxide led astronomers to hypothesize that the surface temperature must be very high, due to the fact that CO_2 is a greenhouse gas.

Eventually, of course, spacecraft did make it to Venus. The first probes to reach the planet came from the United States. *Mariner 2* was launched in December of 1962 and swung past the planet and established that, unlike Earth, Venus had no intrinsic magnetic field. In one flyby, *Mariner 2* destroyed any notions of a tropical paradise. The spacecraft recorded temperatures between 216 °C and 316 °C – far too hot for liquid water to exist on the surface.

Two other Mariner spacecraft also visited Venus. *Mariner 5* flew past the planet in October 1967 and established a surface pressure of nearly a hundred atmospheres. That is a 100 times greater than the pressure on Earth's surface at sea level. *Mariner 10,* en route to Mercury, flew past the planet in February 1974 and photographed Venus in UV light. *Mariner 10* discovered that the wind speeds on Venus were rather high and the photographs revealed some wonderful delicate structure in the ever-present clouds.

It is probably fair to say that it was the Russians who enjoyed the most success in the early robotic exploration of Venus, although they had a number of setbacks. The Russian Venera program was highly successful and really established a lot of new facts about the planet.

Venera 2, alas, failed, but *Venera 3* did manage to reach the planet on March 1, 1966. The plan was that the spacecraft would descend through the thick atmosphere, land on the surface and transmit details from the surface. However, it was not to be. Venus was decidedly more hostile than was thought, and poor *Venera 2* was crushed before it could transmit any data. Still, it does hold the mantle of being the first manmade object to touch the surface of another planet!

Venera 4 arrived in October 1967 and was a little more successful, the spacecraft made a 93 min descent through the dense atmosphere. At a height of some 52 km above the planet, a pressure of 1 atm was recorded and a tolerable temperature of 33 °C. Alas, both pressure and temperature increased, and the probe ceased to function when a temperature of 262 °C and a pressure of 22 atm was reached. A similar fate awaited *Venera 5* and *6*, both crushed by the unforgiving Venusian conditions.

Finally, Russian determination paid off; in December 1970, *Venera 7* became the first spacecraft to land on Venus and transmit back to Earth. It only transmitted for 23 min after landing, but the spacecraft established a surface temperature between 455 °C and 475 °C, and the extensive cloud deck seemed to come to a stop at about 35.4 km above the surface. *Venera 8* touched down in July 1972 and was able to establish that the ambient light on the surface would allow photography.

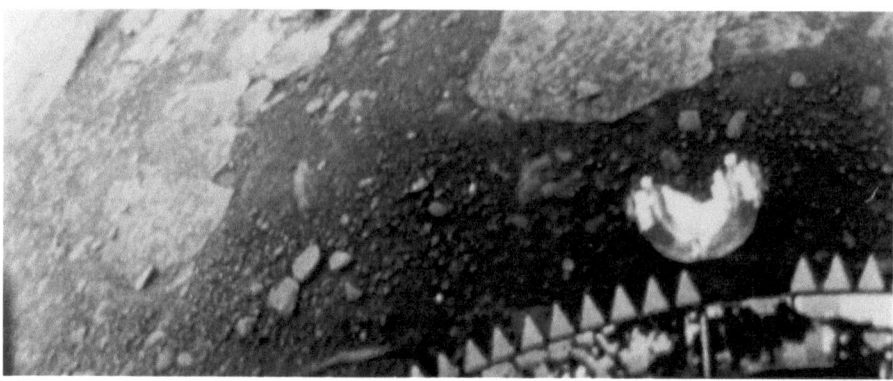

Fig. 4.3 A wonderful colorized image of Venus' surface, seen under a glowing sky

It seemed that the illumination was similar to a very overcast day in winter. The surface, according to *Venera 8,* was similar to granite.

Venera 9 was a great success. It arrived in June 1975 and not only reached the surface safely, but managed to transmit the first-ever images of the surface. *Venera 9* made a 180° panoramic of the surface. Even today, the photographs from *Venera 9* may be some of the best ever taken by a spacecraft, in particular the one shown in Fig. 4.3, showing the baking rocks and the flat landscape forever hidden from Earth by the angry yellow glowing clouds. As well as providing us with a picture of the surface, *Venera 9* found the clouds to be composed of hydrochloric acid, hydrofluoric acid, bromine and iodine. The chief gas in the atmosphere was carbon dioxide. There could be no doubt, Venus was baking, encased in a hot alien atmosphere, harsh and extremely poisonous.

The Venera mission continued up until *Venera 16* (which, together with *Venera 15,* was concerned with mapping the poles of the planet), after which time the Americans had caught up and on May 4, 1989, launched their *Magellan* spacecraft. The principle task of *Magellan* was to map the surface of Venus using radar. This it did, producing a high resolution map of the surface.

Magellan covered 98 % of the surface (Fig. 4.4) and revealed some very interesting facts. For a start, the surface appeared to be covered with materials ejected from volcanoes. There were mountain ranges, two large highland areas (Ishtar Terra in the north, and the Aphrodite region that starts near the equator and crosses south). There are also many extensive channels cut by slow flowing lava rather than water. Some of these channels were nearly 6,000 km in length.

The surface is also covered with large pancake-like structures that may well be some sort of large volcanic domes. Vast concentric radial tracts called arachnoids (as they resemble spider's web) can also be found on the surface. One interesting fact that turned up in the survey is the age of the surface of Venus. We can measure the age of a surface by looking at the impact craters.

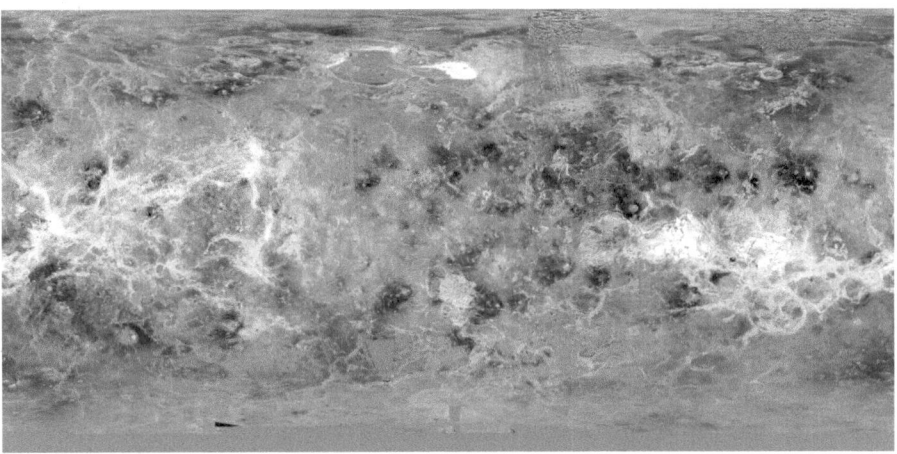

Fig. 4.4 *Magellan* map of Venus

Of course only large impactors are going to make it through that dense atmosphere, those capable of generating craters of 30 km or more. It seems that Venus has rather few, which makes its surface fairly young, maybe even less than 800 Ma old. Moreover, the random distribution of craters over the surface implies that the whole surface is about the same age. This is very unusual. Here on Earth, volcanism makes new land, and on the Moon, we see a surface of different ages. Could it be that Venus resurfaced itself in the recent past?

Due to the relative thickness of the crust, there is no plate tectonic activity, and it is thought that underneath the crust we have a mantle and an iron-rich core. In the terrestrial planets, it is believed that a strong magnetic field such as we have here on Earth comes about due to the metallic core; since there is no magnetic field on Venus we have no definite proof that there is a metallic core at the heart of the planet. However, it is almost impossible to imagine a model of Venus without one.

The arrival of the Space Age allowed us to send spacecraft to pierce the clouds of Venus, and finally to build up a truer picture of the planet and its composition. As is usual in the course of such investigations, more questions were generated than answers. Spacecraft can only spend a limited amount of time examining the planets, and professional astronomers and observatories are often engaged in more fundamental studies, so it is up to the dedicated amateur to keep a scientific vigil of the planet and keep abreast of any recent changes. Fortunately, amateurs have been watching the planet for a long time. However not everything that our devoted ancestors observed offered a quick and simple explanation.

Historical Accounts and Observations

Being the brightest of the main five bright planets, Venus would have been visible to the ancients. What did early prehistoric humans make of this sparkling visitor to their prehistoric skies? The planet is recorded in the *Venus Tablet of Ammisaduqa,* dated 1581 B.C., and it is clear that the Babylonians realized that the bright morning star and the bright evening star were in fact the same object.

Naturally, it was one of the first objects to be examined telescopically, and the first recorded Venus observations come from Galileo Galilee. On December 11, 1610, Galileo was able to discern the phase of Venus (indeed, the phase can be seen in a pair of steady binoculars), but due to his small aperture telescope, there was little more that could be done.

In 1666 and 1667 J. D. Cassini observed the planet and several white spots on the disk, and in 1726 and 1727 Bianchini repeatedly observed some dark spots on the disk of Venus while he was in Rome. So distinct were the outlines of these spots that Bianchini believed them to be fixed structures, genuine markings on the surface. He even charted and named them. When it was revealed that Cassini, who was using a larger telescope in Paris, could not see Bianchini's markings, the proposed solution was that Paris was an unsuitable location to view these delicate features!

Moving on into the late 1700s, we find Shröter's observations of the planet. He observed the planet in 1788, and for many years he observed nothing but the fall off in light at the terminator, which he correctly attributed to an atmosphere (see Fig. 4.6 later). However he failed to register the bright cusps' caps. He did, however, observe something else. On 1789, December 28, he turned his telescope towards the planet, and noticed something unusual. He believed he had seen the southern horn of the crescent truncated with an isolated bright point at its extremity. It reminded him of what he often saw on the moon. Shröter toyed with the idea that it might be a mountain, but if it was, it would be nearly 22 miles high – four times higher than the mountains of Earth. Suffice it to say it was a visual fallacy, but it had important consequences. The observation and other subsequent observations gave rise to the 'Himalayas of Venus'. For more on this and its important consequences see Richard Baum's superb historical book *The Haunted Observatory* (New York; Prometheus, 2007), which tells the story in great detail.

It was this observation that Shröter followed through until the planet became a thin crescent with extended cusps at inferior conjunction, which brought about the recognition of twilight on the planet. This fact gave astronomers what they had long sought – irrefutable proof the planet has an extensive atmosphere.

It is very easy for us to sit here and disparage the observations and conclusions of those early explorers. Thanks to spacecraft, we can see the features of the planets without even using a telescope; indeed we do not even need to be astronomers to explore the worlds of our solar system. But that course of action is as profitless as it is unwise. Shröter did exactly what we do in science today. He observed something unusual and tried to explain it based on what he knew and had seen already.

Shröter had viewed the Moon, and even in a low power eyepiece, when the Moon is not full, the terminator is not a straight line. Rather it weaves a fine and delicate path around crater edges and long mountaintops. The dramatic geology on the Moon means that the terminator evolves over the course of a night. The tops of the mountains in the southern uplands glisten and shine like tiny beads of light in a pool of ink – the result of sunlight hitting the mountain tops and over the course of time illuminating the whole mountain. Shröter would have observed this many times on the Moon, so if he observed an irregular terminator, and a point of light on Venus, and was confident his telescope was not out of alignment, what else was he to conclude? In the absence of knowledge, the first faltering steps towards the truth have to be made by someone. Shröter's Himalayas do not literally exist; they were a mistake, but a honest one.

Further Accounts and Observations

Other astronomers began to examine the planet in earnest. Madler observed the planet from 1833 onwards. He observed a number of brighter regions on the disk and irregularities in the terminator and cusps. Attempts were made at finding Venus' rotational period: In 1666–1667 Cassini estimated it was 23 hours 21 min. In 1726 Bianchini arrived at the number of 24 days 8 hours for the rotation. In 1811, Shröter obtained a value of 23 hours 21 m 8 s, and in 1840–1841, Di Vico obtained a similar value to Cassini's. Alas, all of the figures are wrong, simply because the thick clouds of Venus never clear away and no permanent feature can ever be glimpsed on its surface. We had to wait until the twentieth century before the true rotation of the planet was revealed.

One thing that is obvious when we observe Venus in a telescope are the bright polar regions. Thanks to spacecraft we know that these are bright swirling vortices, but back in previous centuries the people believed the bright poles to be evidence of polar ice caps. Again, based on the evidence at the time, this was not unreasonable; the poles of our own planet are covered in snow and ice, and the poles of Mars, too, have polar snows that come and go with the changing seasons.

Probably the first person to remark on the bright polar caps was the German astronomer Franz von Paula Gruithusien, born on 1774, March 19, at Haltenberg Castle near Kaufering. Gruithuisen seems to be to have been a rather unusual character with some different ideas. He believed he had observed cities on the Moon, and advanced the theory that the Ashen Light (see next section) was the result of Venusians celebrating with fireworks and city lights. Although he had a active imagination, he did make a number of reasonable observations of Venus.

On the 1813, December 29, in the early evening, Gruithusen was examining the planet when he noticed the southern polar regions were different from before. He changed telescope, and using a power of ×135, he observed two very small bright spots near the southern horn. He observed a similar spot near the north pole. Then, at around 1700, the north pole appeared to get brighter, and the southern polar spots had merged into one bright spot. He observed the southern spot again on December 30 and 31. A single spot was visible on January 14, but by 1701, February 21, the hood appeared to be "divided into two small distinct points of light." Gruithusen assumed that the poles of Venus were covered in ice and snow, and the smaller points of light were small islands of ice that circled the poles.

Although Gruithusen's ideas were supported, later on it was Schiaparelli and Antoniadi who suggested that the idea of the polar hoods being the result of snow and ice was wrong, and in fact they were just bright clouds.

Interestingly, unusual disk markings continued to be seen well into the twentieth century. Henry McEwen observed sparkling stars in the planet, and a bright limb projection was observed by Alan Heath on 1958, January 19. Sir Patrick Moore and many others have often observed and photographed a crescent with blunted horns. What are we to make of this? One of the tasks of the visual observer should be to try to obtain repeat observations of these unusual phenomena. It is either the case that these observations are the result of less than ideal optics, or occasionally the atmosphere of Venus produces some surprises.

The Ashen Light

Without a doubt, one of the main enduring observational mysteries associated with Venus is a phenomena called the Ashen Light. The light normally takes the form of a greenish-gray light that can be seen on the dark side of the planet, normally when the planet is in a crescent phase. It was first observed by Johannes Riccioli in 1643. Many Earth-based observers have since reported the phenomenon. What could this ghost light be? One explanation is that the atmosphere of Venus occasionally thins, and this allows the hot surface to shine through. Of course the problem is that this IR transmission would probably be undetectable with the naked eye, and it's probably safe to say, if the effect is truly a physical one, the mechanism that produces it is still very much a mystery.

In Fig. 4.5 are some drawings of the phenomena made by reliable observers of the planet.

Observing Venus

It is no secret that Venus is a difficult world to observe. It never climbs too high above the horizon, and is only ever an evening or a morning object. Nevertheless, if we adopt careful observing practices, it is possible to do a great deal of useful work.

We can choose to observe Venus under two circumstances, either during daylight hours or during the early morning or late evening when the planet is a bright object in the skies.

Observing in daylight has advantages; for one thing the planet is high in the sky, and so is much less affected by poor seeing. The glare of the planet is also

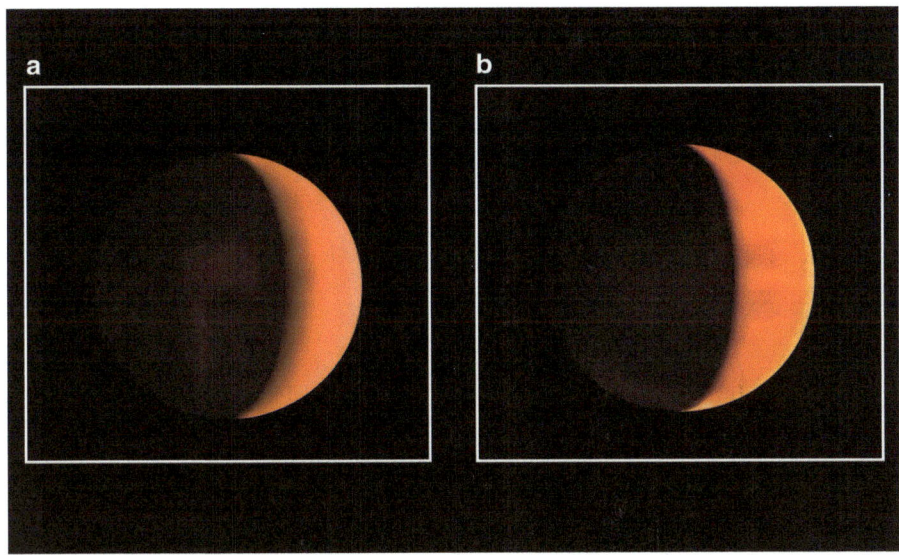

Fig. 4.5 Two drawings of the Ashen Light by David Gray: (**a**) 2007, September 23, at 0510UT, ×365 with a W#22 filter, Seeing AII-III, and (**b**) 2007, September 26, at 0500UT, ×365 with a W#22 filter, seeing AII. Both drawings made with his 415 mm Dall-Kirkham (Courtesy David Gray)

considerably reduced. The main downfall of observing in the day is of course that you have to be very keen sighted indeed! Many observers use setting circles (i.e., the R. A. and dec dials on their telescopes) to find the planet in these conditions.

I should interject a word of warning here. If you are trying to observe Venus during the day, be very careful of the Sun! Even a second's glimpse in the viewfinder is enough to do a lifetime's worth of damage! If you are inexperienced, or unsure of the planet's location, it is best to observe the planet when it is in the early morning or evening sky.

Although the planet will be quite low down, there will be no difficulty in finding it. To cut down the glare an apodizing screen is most helpful, and of course filter work can be undertaken due to the brilliance of the planet.

As you will no doubt discover, Venus does not tolerate high magnification well, and a power of 160×–200× is generally more than enough. When observing the planet, you will find it helpful to experiment with the magnification. Use a power that matches the conditions. It may well be the case that your telescope can go up to 400×, but if all you get in the eyepiece is a wobbly steam pudding, then there will be little point in using such high powers.

Most people who spare Venus a quick glance through a telescope will see very little (or indeed nothing). It is true that the cloud markings are delicate, but they are often faint and elusive, and it may well take you a good 15 min before you see

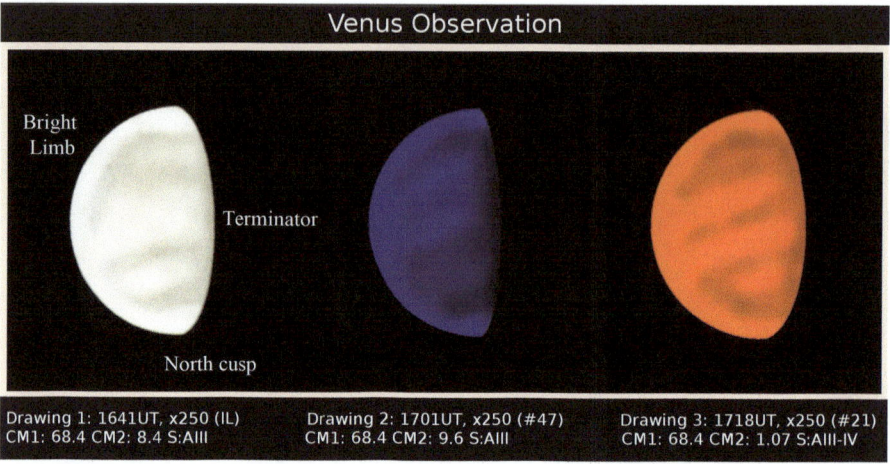

Fig. 4.6 Three drawing of Venus made by the author on 2012, January 10, at the times indicated with a 203 mm Newtonian Reflector. Drawings were made in IL, and with a *violet* W#47 filter and a *orange* W#21 filter. The IL drawing has the bright limb band, cusps and terminator marked on

anything definite. In Fig. 4.6, I have given the general features you will see in a 3 in. (90 mm), or larger telescope.

First of all, we have the whole illuminated disk. This will not be the whole disk, and there will be a phase depending on where in its orbit Venus lies. In general there will be faint grayish markings on the disk, along with brighter and darker regions. You may also catch glimpses of the various "psi" shaped markings.

The next thing we have is the terminator, which crosses the planet from north to south and is the line between night and day on the planet. Normally the terminator is smooth, but occasionally it can appear irregular, so it is worth examining the terminator to make an assessment of its appearance.

Opposite the terminator, and extending from the north to the south along the edge of the disk we have the bright limb band. This is normally the brightest feature on the disk and can be very dazzling indeed. Sometimes, however, the limb band may not be complete. The dusky marking on the disk may encroach upon it or some sections of it may be brighter than others.

The cusp regions can be found covering the poles of the planet. They are usually the same brightness, but sometimes one cusp may be brighter than the other, or one of the cusps may be absent. Bordering each of the cusps we have the dusky cusp collars, and, like the cusps, they may not be of equal intensity or always present. When you first have Venus in the telescope, it is worth checking on all these features for about 10 or 15 min. Once you are certain you have a picture of what's there, it is time to make a drawing.

Drawing Venus

Making a drawing of Venus is a relatively simple task, but there are few pitfalls awaiting the unwary observer. First of all, observe with a circular blank 50 mm in diameter. Now, you may be tempted to look up the phase of the planet, and draw this onto the blank before you go out. This is a mistake. There is a phenomena on Venus known as the Shröter effect – the thick atmosphere of the planet scatters the sunlight, so the predicted phase is usually different from the actual observed phase! Thus dichotomy may be predicted on a certain date, but in practice, the planet will appear to be about half a few days before or after then.

Once you have your blank and are ready to make a drawing, start by putting in the line of the terminator as accurately as you can. If there is any shading near the terminator, put this in, too. Next, fill in any shading on the disk, including the cusp caps and collars. You might want to emphasize any bright regions on the disk, or where the bright limb band meets the general hue of the disk. Finally, note down the time of observation (in UT), telescope and magnification used, along with the seeing conditions. You might want to annotate your drawing indicating any brighter regions, or cloud markings that appear to be unusually dark, etc.

One other thing: always make sure you include the CM longitudes with your drawings, as this will allow you to keep track of interesting features. Venus has two systems of longitude: CM1, which is the longitude of the atmosphere, and CM2 the longitude of the surface. We have two systems of longitude because at visual wavelengths, we can only ever see the cloud deck. You can get CM1 and CM2 from *WINJUPOS* simply by inputting the date and time of your observation.

When you have finished observing, it is always nice to make a final neat drawing. In order to do this, you will need to measure the phase. If you use a 50 mm blank you can do this by first measuring from the distance along the disk center from the limb (Fig. 4.7), to the furthest point on the terminator. If this distance is X mm then the measured phase is:

$$Ph_{measured} = \frac{X}{50}$$

You can then use an appropriate blank with the phase already printed on the disk. Once you have made this drawing, it is best to put it in a notebook for your neat drawings.

Like Sir Patrick Moore, you might want to have observing books for Venus, Mars, Jupiter, Saturn and Uranus that contain color drawings and notes summarizing the important points of the observation and the data you have collected. Quite apart from anything else, having an extensive visual record will allow you to see how you have developed both in observation and draftsmanship over the years!

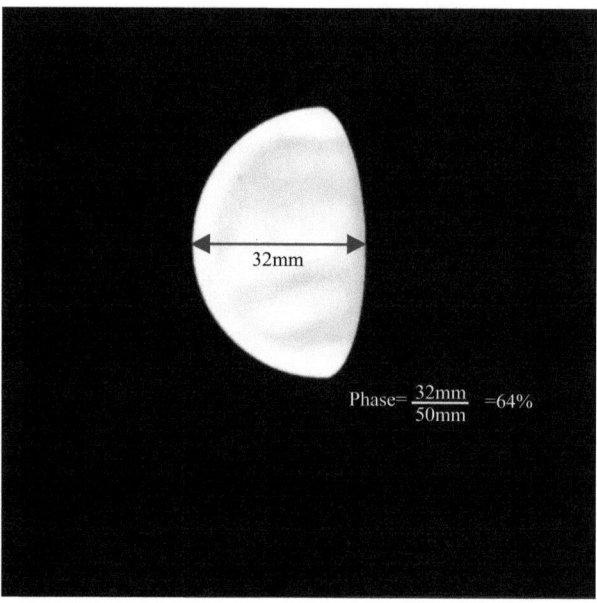

Fig. 4.7 Estimating the phase from an observation

Advanced Projects

Disk drawings are not the only useful thing the amateur visual observer can carry out. There are a number of other projects that are well worth embarking on. Let's have a closer look at some of them.

Intensity Estimates

It is useful to keep track of the intensity of the fleeting clouds and the permanent features such as the limb band and cusp caps. This has been a long-term project of the BAA for many decades now. Making an intensity estimate is a simple affair. You simply make another disk drawing then assign the various features a number between 0 and 5: 0 = extremely bright features, 1 = bright areas, 2 = general hue of disk, 3 = shading near the limit of visibility, 4 = shading well seen, and 5 = extremely dark unusual shading. You should keep a copy of your intensity estimate disk drawing with your neat Venus drawing, and then you can see how the dark and bright features have developed over the course of an elongation. Such observations provide clues about the changes in the planet's atmosphere.

Estimation of Phase

As mentioned earlier, there is always a difference between the theoretical predicated phase of Venus and the observed phase due to the Shröter effect. It is worth plotting both the predicted phase (which you can again get from *WINJUPOS*) and the observed phase on a graph. All you have to do is put the date on the bottom axis and the phase value running along the vertical axis.

Again such data can be used to help confirm if there are any atmospheric changes taking place; if the atmosphere should thin enough in a detectable way, there will not be a great difference between the theoretical phase and the observed phase. If the atmosphere remains the same, the respective curves between the predicted phase points and the theoretical phase points will remain a constant distance from each other.

Filter Work

There is much work to be done here, especially if you own a range of decent filters. For Venus, the following filters are recommended:

- W#47 (Violet). This filter is very useful for studying the upper atmosphere of Venus; a 3 % transmission #47 will block all but 3 % of the incoming light, so you will need an 8-in. or larger telescope to get the most out it.
- W#12 (Yellow). This filter contrasts sharply with blue-green wavelengths of light, and so it should help increase the contrast of any bluish features on the disk.
- W#21 (Orange). This should also help increase the contrast between any bluish-green regions that may be present on the disk.

It is very useful to make disk drawings, intensity estimates and phase estimates using various different filters, as the features and phase will almost certainly appear differently at different wavelengths of light. By doing this you are extending your study of the planet from ordinary white light (or integrated light, as it's often called) to a wider spectrum of wavelengths. Should any changes occur on the planet, you may well be one of the first to notice if you undertake the extensive observing program we have looked at here.

Venus may have a reputation as a awkward customer with a bland disk, but after a few elongations you will soon learn that this is a myth. As with so many things, all Venus really needs is careful attention and a familiarity with what has gone before.

Chapter 5

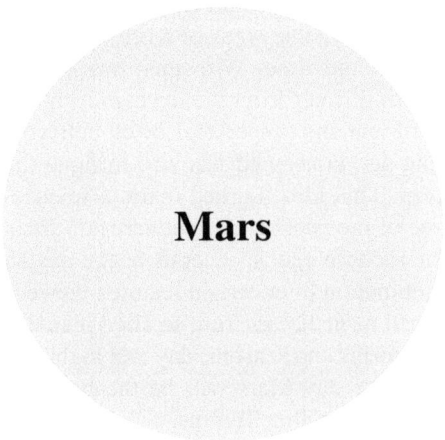

Mars

Useful Facts

Distance from the Sun: 249,100,000 km (max), 206,700,000 km (min)
Diameter: 6,794 km
Sidereal Period: 686.98 days
Synodic Period: 779.9 days
Rotational Period: 24 h 37m 22.6 s
Surface Temperature: −137 °C to +26 °C
Apparent Diameter: 25.7″ (max.), 3.3″ (min.)
Albedo: 0.16
Brightest Apparent Magnitude: −2.8
Satellites: Two

A Brief Look at Mars

Whether as 'Barsoom' of Edgar Rice Burroughs or 'Malacandra' of C. S. Lewis, Mars is the planet that has had the most marked effect on the imagination. A biennial visitor to our skies, named by the Romans after the god of war, because of its baleful luster. It mystified the earliest astronomers by its failure to obey their geocentric idea. Even the Copernican model failed to accommodate its motions, until the observations of Tycho Brahe and the mathematical genius of Kepler finally established that Mars moves in an elliptical orbit around the Sun.

P.G. Abel, *Visual Lunar and Planetary Astronomy*, The Patrick Moore
Practical Astronomy Series, DOI 10.1007/978-1-4614-7019-9_5,
© Springer Science+Business Media New York 2013

Moving on into the telescope era, Mars continued to puzzle. Some astronomers claimed its surface was crisscrossed by a network of fine lines, though others were not convinced. If they existed, what could they be? To the American astronomer Percival Lowell they were the visible signs of a brilliant civilization caught up in a tragic battle against nature and time. With their world dying, they covered the planet in canals, transporting water from the ice caps, down to the parched deserts near the equator. How brilliant and desperate, Lowell believed, were these creatures who had, no doubt, long ago consigned war and intrigue into their past, survival being their main concern. This idea sparked many a science fiction yarn, making the Red Planet the home of monsters and nightmares of dreams and illusions.

As we move into the modern age, spacecraft of the day showed the canals had been a cruel trick, a combination of uncertain features viewed through the unsteady air of Earth. The spacecraft turned Mars from an alien planet into a world. Mars had mountains and canyons, storms and seasons, day and night. How similar to our own world it seemed. Surely, one day Mars will be the home to human beings who might call themselves Martians rather Terrans.

Today, Mars still continues to fascinate. Recent evidence shows that liquid water may exist beneath the surface. Our hopes of finding life on Mars have once again come to the fore. It is relatively obvious to scientists now that Mars was once warm and wet, and may well have internal water sources, the dried up river beds and ancient sea floors might contain the fossilized remains of any ancient Martian lifeforms.

With so many spacecraft and rovers on Mars it might be thought that the telescopic era was done with, and in any case, little more could be accomplished by amateurs. Of course, this is completely wrong. The spacecraft give us a localized view; for a global picture the amateur still has a valued place, often discovering dust storms and changes, monitoring the coming and going of the polar ice caps and so on. There is much work to be done connecting the 'what was' with the 'what is' and predicting the 'what will be.'

Let us now begin our own exploration of the Red Planet, and see what past observations have to tell us about how to make good quality observations in the future.

Orbit and Apparitions

Of all the five bright planets (Mercury, Venus, Mars, Jupiter and Saturn), it was the motion of Mars that baffled the ancients the most. Mars appears to speed up, slow down, loop back on itself and start to move backwards. It was only when Kepler established his laws of planetary motion using the fine observations of Tycho Brahe that the reason for this bizarre dance became clear. Mars is further out from us and so we catch it up in its orbit, then move past it. The drawing in Fig. 5.1 illustrates this.

In Position 1, Earth is behind Mars but starting to catch up to it in its orbit. As we pass through Positions 2 and 3 and Mars appears to be moving quite quickly in the night sky, we overtake the planet, and from our point of view, the planet appears to move backwards (because we are overtaking it). At Position 4, the planet is said to be at opposition; this means that the Sun, Earth and Mars lie on a straight line, from

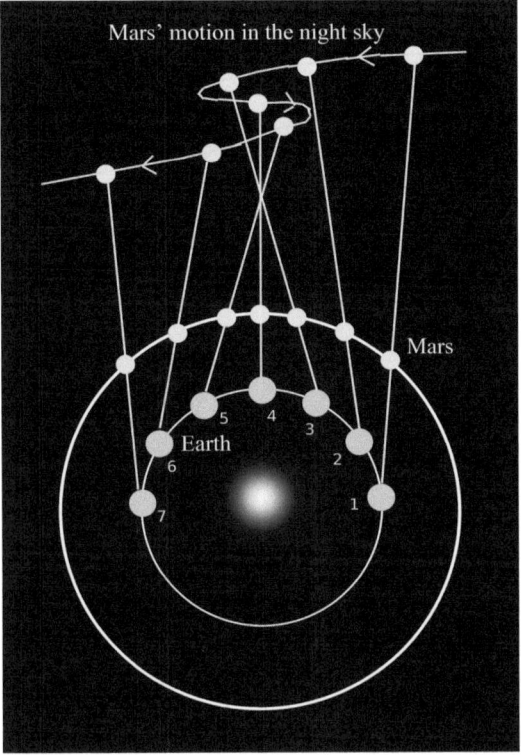

Fig. 5.1 Orbit and motion of Mars in the night sky

our point of view. Mars is in the opposite part of the sky to the Sun; it is due south at midnight and visible all night. We then overtake Mars in Positions 6 and 7, and move away from the planet.

Mars' orbit is decidedly elliptical, and this has two effects. First it means that the planet reaches opposition approximately every 2 years (779.9 days to be exact, called the synodic period). Not every opposition is equally favorable. In fact there are two types of apparition, an aphelic opposition and a perihelic opposition. As Mars has an elliptical orbit, this means there will be a point in the orbit when it is at its closest to the Sun, and this is called perihelion. There will also be a point in the orbit when it is furthest from the Sun, and this is called aphelion. Obviously, if a planet has a circular orbit, it is always the same distance from the Sun.

An aphelic opposition (see Fig. 5.2a) occurs when Mars reaches opposition when it is at Aphelion. As it is far away, the apparent diameter of the planet is small; often it never exceeds 14″. For northern hemisphere observers though, the planet will be high in the sky. It will pass overhead from UK latitudes and so it will not be plagued by poor seeing and unsteady air. By contrast, for the southern hemisphere the planet will barely be above the horizon, it will be lost in the murk and

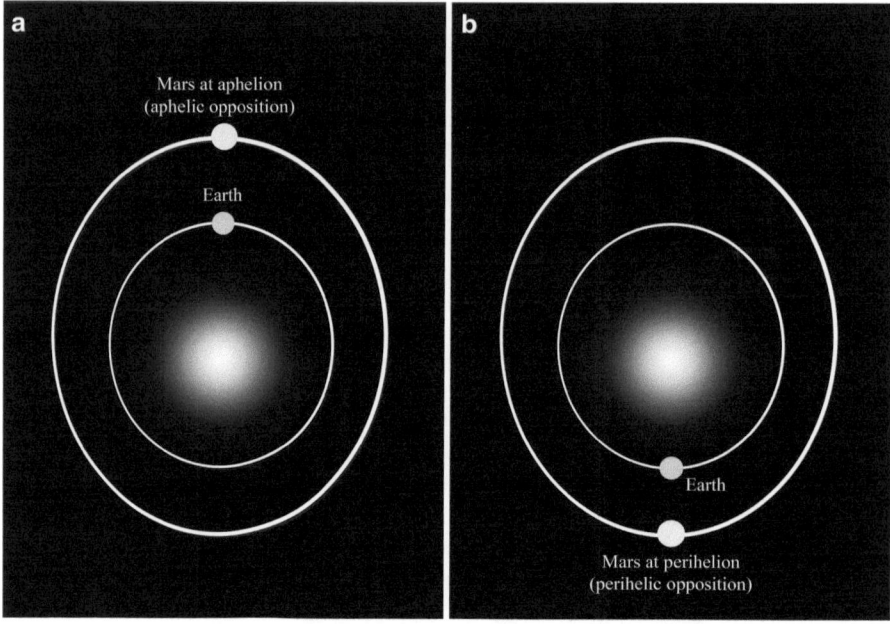

Fig. 5.2 (a) Mars at aphelic opposition, when it is furthest from the Sun. (b) Mars at perihelic opposition, when it is closest to the Sun

turbulent air, and that combined with the small disk size will make the apparition largely a write off!

Perihelic opposition occurs when Mars is at perihelion (see Fig. 5.2b) and the tables are turned" Mars is closer to us in its orbit now, and the disk size can be as much as 26″. In the southern hemisphere, the planet swings overhead, and the southern people are afforded a magnificent view of the planet. Alas for northern hemisphere observers, the planet is low down, for some UK latitudes, it will barely scrape above the horizon and will be blocked by houses and trees. In any case, the planet will be so low down, even on a clear horizon, the fine features will be hidden by the thick turbulent air of our atmosphere.

Oppositions cycle from aphelic to perihelic and back again, and clearly then, for northern hemisphere observers, aphelic oppositions are best. For southern hemisphere observers, perihelic oppositions are favored. Both hemispheres get a reasonable view at those oppositions between aphelic and perihelic ones.

Like Earth, Mars has an axial tilt, and this means its northern and southern hemispheres experience the same familiar seasons we have here on Earth – winter, spring, summer and autumn. Now, it is very useful to keep track of Martian seasons, as there will be various phenomena associated with each one (like the thawing of polar ice caps). To enable us to keep a seasonal track of Mars, astronomers have devised a quantity called Martian solar longitude, and we denote this quantity L_s. In Fig. 5.3 is a drawing of how Ls works.

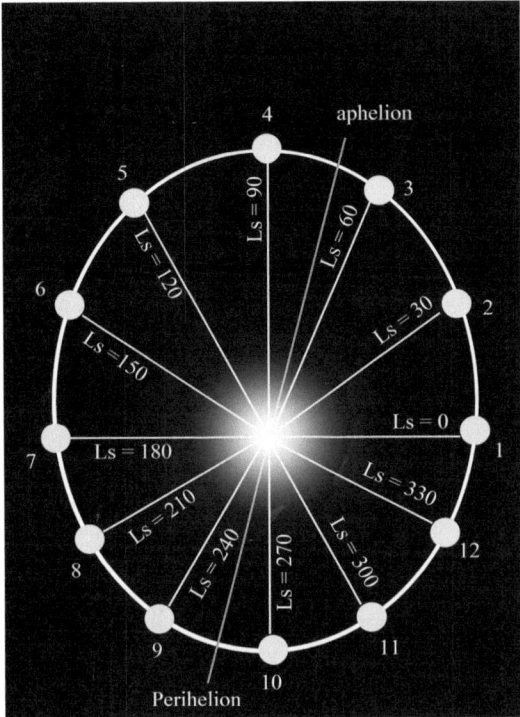

Fig. 5.3 Drawing showing Martian solar longitude over the 12 months of the Martian year

In Fig. 5.3, we have a drawing showing Mars in its orbit around the Sun. It takes Mars 686.9 Earth days to go around the Sun once. We have four main points: Ls=0°, 90°, 180°, 270°, and so each Martian season is 30° of Martian solar longitude. Due to the elliptical orbit, not all seasons are of equal length. Let's see what happens in the northern hemisphere of Mars during the changes in *Ls* (for southern hemisphere, simply reverse this as spring in the north corresponds to autumn in the south, and so on).

We start then with Mars at Position 1, which we shall call Month 1, and this corresponds to *Ls*=0°. This point is the spring equinox, and so Months 1, 2 and 3 are springtime. This means the northern polar ice cap will be thawing and returning quantities of water vapor back into the atmosphere. Over these 3 months, the cap will get brighter as the ice melts and bright fogs and mists are created. Eventually the cap will recede as the northern hemisphere becomes warmer. This is a good time to look for bright orographic clouds. During Month 3, Mars reaches aphelion (at *Ls*=71°) and is the furthest from the Sun it can ever be.

We now come to Month 4. The point *Ls*=90° is the summer solstice, and this marks the onset of summer. By now, the north polar ice cap has retreated, and from Earth the cap appears to be very small indeed. We can now see the regions that had previously been covered in frost and ice. Bright clouds may still be observed in the north.

Moving on to $Ls = 180°$, we come to the autumn equinox. It is now starting to get colder, and so the ice cap will start to reform and increase in size. This point also marks the start of the Martian dust storm season. We can now expect the winds of Mars to produce dust storms, regional ones affecting smaller areas, or powerful ones whereby a vast amount of dust is swept into the thin atmosphere, and all the features of Mars may be hidden for months! The dust storm season will last from Months 8–12, and the resulting storms may leave the various permanent features on the surface changed.

Mars reaches its closest point to the Sun (perihelion) at $Ls = 251°$. We next come to position, $Ls = 270°$, or the winter solstice. By now the northern polar cap has advanced, and it is now large and very bright. The dust storm season has still been continuing, and if a global dust storm has struck, all that may be visible from Earth is the bright polar cap and a rather blank orange disk. We continue on back to Months 11 and 12, at around $Ls = 330$, the Martian dust storm is over, and soon spring will start. After this the whole cycle will repeat once more. You can find out what the current value of Ls is by using *WINJUPOS*. Simply go to the ephemerides page and look up the value called longitude of the Sun. This is the current value of Ls.

As can be seen in Fig. 5.3, at aphelic oppositions we get to see spring and summer in the northern Martian hemisphere, which is tilted towards us. As Mars is far away, its apparent disk size from Earth is quite small. Conversely, during perihelic oppositions, it is autumn and winter in the Martian northern hemisphere. However, it is the southern Martian hemisphere that is tilted towards us, and we get to see spring and summer in the south. The southern cap retreats and, due to the large disk size, we are afforded some wonderful views of the Red Planet.

Let us now move on to look at the planet itself, and what we currently know and understand the planet to be.

The Planet Mars

Of all the planets in the Solar System, Mars is perhaps the closest to Earth; it is a rocky terrestrial planet whose internal structure is very similar to Earth. Figure 5.4 shows a cutaway of the planet, revealing its interior.

The planet is surrounded by a rather thin atmosphere that consists largely of the gas carbon dioxide. Carbon dioxide makes up around 95 % of the atmosphere, the other components being nitrogen and argon. Although the atmosphere is weak, it does support clouds that can produce precipitation; snow has been observed falling by the Phoenix lander. The atmosphere can also produce surprisingly substantial winds of up to 250 mph. Such winds transport vast amounts of the iron oxide dust around, and many of these lighter particles remain in the air. Quite recently, methane was detected in the atmosphere of Mars by the ESA spacecraft Mars Express. This is an exciting development, since methane is normally associated with biological activity. Moreover, the amount of methane seems to be constant, which means it is being replaced. (It should be broken down by ultraviolet radiation in around

Fig. 5.4 Internal structure of Mars

350 years.) Biological activity is not the only possibility, and the explanation for methane production maybe a geological one. At the time of writing no consensus has yet been reached as to the cause.

The thin atmosphere makes the atmospheric pressure very low and could never support liquid water on the surface. Even at the bottom of the deep Hellas Basin, the pressure is little more than 8.9 mbar, and at heights of Olympus Mons, it can be as low as 6 mbar (compare this with the pressure on Earth, which, at sea level is 1,013 mbar). The Martian atmosphere is also poor at retaining heat received from the Sun. Here on Earth, our atmosphere traps heat and keeps the surface of the planet at stable temperatures. On Mars, the temperature in the summer in the southern hemisphere may reach as high as 26 °C. The lowest temperature can be found at the poles and is a harsh −137 °C, much too cold to support life. Various spacecraft have measured the temperature at the surface, and the average temperature seems to be around −55 °C.

The surface of Mars is composed largely of igneous rock, and seems to be largely cold lava. In many places the surface is covered in soil that contains large amounts of iron oxide (rust), which gives Mars its distinctive reddish hue. Magnesium and potassium have also been detected. The surface is just the top if the crust, and it seems the Martian crust may not be uniform, and may be substantially thicker in some places (like the Tharsis volcano complex) than in others.

Beneath the crust we come to the mantle. This region is most likely composed of various silicates. Studies of Martian meteorites suggest that there may be twice as much iron in the mantle of Mars than that of Earth. The core of Mars is also likely to be composed of iron silicates and sulfur. On Earth, our own mantle and core are molten; the mantle is the lava material produced from erupting volcanoes. Earth's magnetic field is driven by a dynamo action in the molten core.

On Mars, however, volcanic activity seems to have stopped many millions of years ago. Moreover, the planet's magnetic field is very weak. This would seem to suggest that the interior of the planet is much colder than Earth. This is to be expected since Mars is much smaller than Earth and would have radiated its internal heat away much more quickly than Earth. As you might expect, we know very few definite facts about the interior, and the debate on the exact internal structure of the planet is likely to continue without resolution for some time yet.

Mars in the Telescope

As we have seen, Mars has a rather thin atmosphere, and unless there is a global dust storm in place, the surface displays a wide range of markings known as albedo markings. These markings are vast plains and lofty plateaus, and you will find it enormously helpful, if you can find your way around the various features on the planet. At first this may seem like a daunting task, but it is really no more than learning your way around a big city. You start with the big striking features and move on from there. In this section, we shall explore the surface as seen through the telescope at intervals of 60° in longitude.

Zero longitude on Earth is to be found at the Greenwich meridian. Essentially longitude tells you how far east or west you are. So the same is true for Mars, although the line of zero longitude is found at the Sinus Sabaeus region. Figure 5.5 is a map made of the planet during the 2011–2012 apparition, also including lines of longitude. The line of zero longitude to 60° seems like a sensible place to start our exploration of the surface. As most astronomical telescopes invert an image, we shall start in the south and move north.

Longitude 0° to 60°: Acidalium, Chryse and Erythraeum

This region is dominated by two dark albedo markings; Mare Acidalium in the north and Mare Erythraeum in the south. Starting in the far north, Acidalium is usually quite a dark feature. When it is spring in the northern hemisphere, and the NPC is retreating, Acidalium extends down into the cap itself. Acidalium is topped by the curving feature known as Niliacus Lacus, which can be quite dark and prominent.

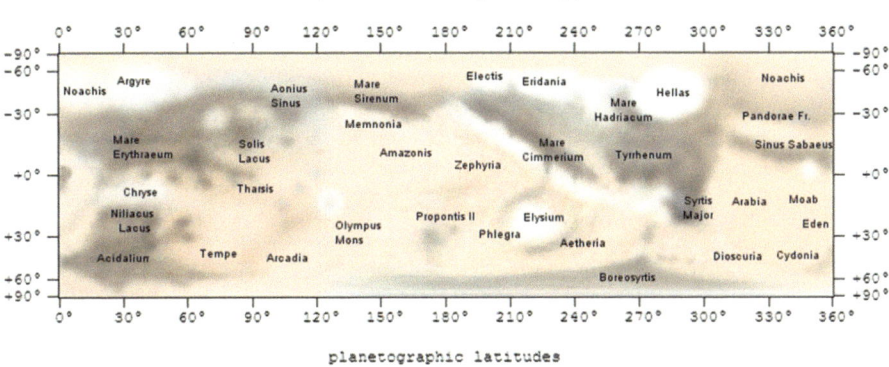

Paul G. Abel

Based on visual observations made using:
203mm Newtonian (Knighton Observatory)
508mm DK (University of Leicester Observatory)

Note: Very few observations were
made covering CM: 140-190

Fig. 5.5 A map of Mars based on the author's observations of the planet in the 2011–2012 apparition using a 203 mm Newtonian reflector

The equatorial region of this part of Mars is marked by the bright area known as Chryse. The region is a basin, and spacecraft have revealed extensive water erosion and what appear to be outflow channels. This region has provided a wealth of evidence that Mars was once warm and wet. Chryse itself may have been a small sea. One must wonder when looking at this region if, millions of years ago, life developed and flourished in that now vanished sea.

Moving southwards we have the dark Mare Erythraeum. This feature, like Acidalium, can be seen in quite small apertures when the planet appears sufficiently large. Erythraeum is another dark plane, but it is not uniform in appearance. There are darker and lighter regions, and when the seeing is good, at a power of ×350 the whole region shows some fantastic fine details!

Above Erythraeum, in the far southern uplands we have the impressive Argyre Basin. This is the second largest impact basin on Mars (Hellas being the first); it extends some 1,120 miles across and is around 3.2 miles deep. Interestingly, the basin floor appears to contain a sort of layered material that points to a sediment laid down at the bottom of a lake or small sea. Was the Argyre impact basin once a small ocean? Argyre normally takes the appearance of a bright ochre patch above Erythraeum, but sometimes the basin fills with white clouds. When this happens the whole of the southern uplands can look ablaze, the clouds being so bright. The basin has also been the site of dust storms (both local and regional) in the past, so the area is well worth keeping an eye on!

MOC2-134a   Malin Space Science Systems/NASA

Fig. 5.6 Clouds collecting around the vast Tharsis volcanoes (Courtesy NASA JPL Mars Global Surveyor spacecraft – mars.jpl.nasa.gov/mgs/msss/camera/images/6_10_99_tharsis/)

Longitude 60° to 120°: The Solis Lacus and Tharsis Volcano Region

This is a fascinating region sculpted by powerful geological forces and time. In the far north we have the bright Arcadia region. In the telescope it appears as a bright orange patch; it is, in fact, a smooth plane formed by lava. Spacecraft have revealed evidence of ground ice near the surface.

Moving south we come to one of the most geologically impressive regions of the Solar System: the Tharsis region. Tharsis contains four large shield volcanoes, the Tharsis Montes trio Arsia Mons, Pavonis Mons and Ascraeus Mons, and further west on the edge of the region, Olympus Mons, the largest volcano in the Solar System. Olympus Mons is visible in many amateur telescopes, while the other three require rather larger instruments to see them. Nonetheless, Martian meteorology gives them away. In the Martian spring and summer, bright white clouds collect around the Tharsis trio creating distinctive "w" or "m" shaped clouds. Similarly, bright clouds often collect around the top of Olympus Mons, making the region very bright and visible in medium-sized telescopes (Fig. 5.6).

We might ask how is it some of the largest volcanoes come to be all in one place. The answer, it seems, is that Mars, unlike Earth, has no plate tectonics. As a result lava is free to build up in one place for millions of years, creating the vast shield volcanoes we see in this region.

Continuing southward we see the start of the great Valles Marineris region. This is a truly colossal valley and makes our own grand canyon look like a trifling trench. It winds around through the Mare Erythraeum and 'empties' into Chryse. The canyon is more than 4,000 km long, 200 km wide and nearly 7 km deep. If you stood on one side of it, you would be unable to see the other side! The surface of Valles Marineris appears to have been affected by running water; indeed it has been suggested that Maris was once a vast river emptying into Chryse. One might wonder, when looking down at this region, did life ever have time to start in these warm and ancient rivers? What kind of life (if any) may have burst forth in these vast channels?

Moving further south we come to a rather famous feature, Solis Lacus, or the Eye of Mars. The view of the region in smaller to medium telescopes, when Mars is close to us, is often of a circular patch with a central dot (giving rise to the name the Eye of Mars); in larger telescopes, under good conditions considerable structure can be seen. Solis Lacus and just to the southwest Aonius sinus appear to be joined by a web of fine lines and points. Over many years the region has demonstrated changes in appearance as dust storms alter it. On a good night at ×400 both the Mare Erythraeum and Solis Lacus appear to be so full of rich fine details it can be difficult to record it!

Longitude 120° to 180°: Mare Sirenum, Amazonis & Propontis I/II

This region is known among amateurs as "the dull face of Mars," due to the lack of strong albedo features. Telescopically the region may appear somewhat bland, but geologically speaking this is a fascinating region of Mars with an abundance of evidence for running water in the past.

Starting in the north we have the lovely ochre desert of Arcadia. To the west we have the vague darker markings of Propontis I and II. These are worth keeping an eye on as they do change their appearance over time (again due to dust storm activity). It is in this region that the largest volcano in the Solar System, Olympus Mons, is actually situated. As we pass through the equator southwards we pass through the Amazonis desert until we reach an interesting region called Memnonia.

Memnonia often appears as a brighter, cream-colored region immediately under the dark Mare Sirenum. Memnonia is a cratered upland. Columbus crater is found here and was the site of exploration by the Mars rover Opportunity. The existence of hydrated minerals in this crater means that it may have contained water in the distant past.

Immediately to the south we come to the most striking feature within this block of longitude – Mare Sirenum (Terra Sirenum). This is a dark and heavily cratered region. The region has been of much interest since the discovery of Martian gullies in Sirenum. Put simply, gullies resemble ditches or small valleys that appear to have been formed by running water. They normally form on hillsides (or on crater walls). Many of the Martian gullies appear to have formed on top of sand dunes and seem

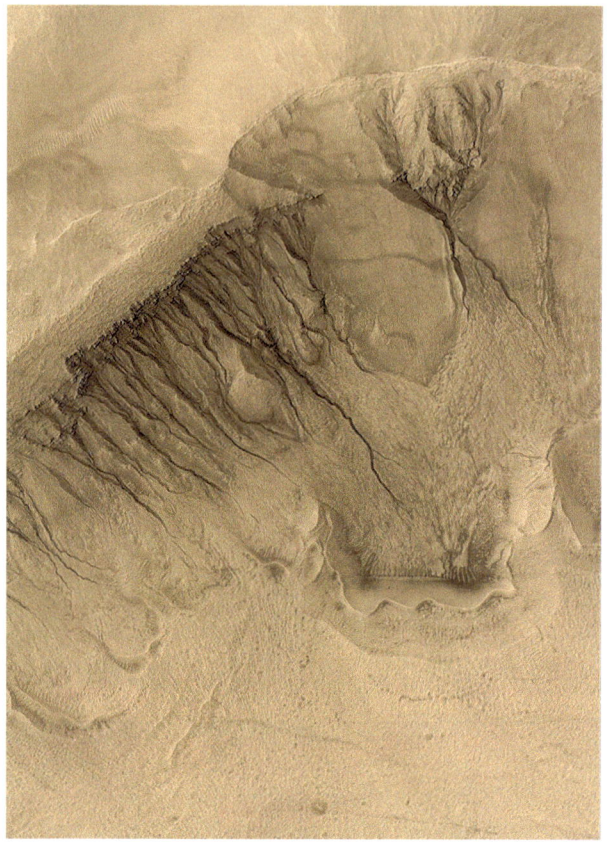

Fig. 5.7 Mars Global Surveyor image of gullies in Newton Basin in Mare Sirenum (Courtesy of NASA JPL– mars.jpl.nasa.gov/msl/gallery/craters/200221007a.html)

to have little cratering, suggesting that they are relatively young. The most likely explanation for their formation is running water coming from the melting of ice of an aquifer (Fig. 5.7).

A further interesting discovery was made by Mars Odyssey orbiter in 1998; this was the existence of chloride deposits within Mare Sirenum dated between 3.5 and 3.9 billion years ago. The existence of such minerals strongly implies that there was a lot of surface water present on Mars in the distant past.

Longitude 180° to 240°: Mare Cimmerium, Eridania & Elysium

This region is dominated by two features. We have the bright Elysium plateau in the north and the dark Mare Cimmerium in the south. In the far north we have the deserts of Cebrenia and Utopia. In the telescope, you would normally find Elysium to be a bright and roughly pentagon-shaped, bordered by the darker Phlegra to the east and Aetheria to the west. Elysium is in fact the second largest volcanic region on the planet. A number of volcanoes have been found here (Hecates Tholus, Albor Tholus and Elysium Mons).

Further south we have the bright deserts of Zephyria, Aeolis and Aethiopis. South of this we have the dark feature Mare Cimmerium (Terra Cimmerium). Like Sirenum, this is another heavily cratered upland region and similarly, Mare Cimmerium is also home to gullies. For many years now faint extensions could be seen running down from the Mare Cimmerium into the brighter Aeolis and Zephyria regions.

In the far south we have two bright regions, Electris and Eridania. Bright clouds often collect in Electris and Eridania. Eridania can become very bright on occasions when there is a lot of cloud activity in the region. Like Cimmerium, Eridania contains a number of gullies and outflow channels, and it is theorized that Eridania may once have been an ancient lake. Near the south pole we have the brighter Thyle II region, which is separated from Electris and Eridania by the darker region known as the Mare Chronium.

Longitude 240° to 300°: Syrtis Major & Hellas

This block of longitude contains perhaps the most distinctive features on the planet. It is dominated by the great Syrtis Major feature, the first object ever recorded on Mars. In the far north we have the Boreosyrtis feature. This is quite a lovely subtle feature that should be visible in medium-sized telescopes when the planet is of a reasonable apparent size. It takes the form of a dusky triangular feature that curves down into the northwest. Near to where it tapers to a point just under Syrtis Major, sometimes it seems that subtle dusky streaks start and flow westwards into the brighter Dioscuria region.

As we move south we come to the impressive Syrtis Major, which in small telescopes looks distinctly 'V' shaped. Syrtis Major was once thought to be a sea, but we now know it is formed by large shield volcanoes. The region is dark, due to the rock being composed of a dark volcanic basalt. There also seems to be a lack of dust in the region. Syrtis Major should be observed carefully because it often undergoes subtle changes after a dust storm, and seasonal changes have also been noted. Syrtis Major also appears to undergo a subtle color change when high altitude clouds are present. When the equatorial cloud band is present, the region seems to take on a slight bluish color. (Indeed, subtle greenish tones have been observed in the region on occasions – no wonder earlier observers thought it to be vegetation!)

As we continue south we come to Mare Tyrrenhum in the east and Iapigia in the west. In medium to large telescopes this region is truly a splendid sight in good conditions. At powers of ×300 or more in good conditions the whole region is filled with fine detail; subtle lighter regions become apparent, so do shadings of varying intensity and color. Tyrrhenum is an old cratered plane, and a large volcano (although it is beyond Earth based telescopes) can be found here – Tyrolean Patera. This is thought to be one of the oldest volcanoes on Mars. Just to the east, the region Libya can sometimes become quite bright.

As we move southwards we come to a truly vast feature, the Hellas Basin. This feature can collect clouds and become very bright (indeed it can become over-whelmingly bight on occasions as the bright clouds reflect the sunlight). When the northern hemisphere is tilted towards us, Hellas, which is then situated at the extreme southern edge of the disk, looks like a brilliant south polar cap. Hellas is an enormous impact basin and is bordered by the dark Mare Hadriacum on its east-ern side. It was probably formed during the Heavy Bombardment period that occurred some 4.1–3.8 billion years ago. This was a violent time in the history of the Solar System, and its fingerprints can be found in the form of craters all over the Moon, Mercury, Venus, Earth and Mars. Hellas is also the site of outflow chan-nels. What must it be like to stand on the tops of the Hellas basin? The view must be truly magnificent.

Longitude 300° to 0°: Sinus Sabaeus, Eden & Cydonia

This region of Mars is dominated by the striking Sinus Sabeus feature. Sinus Sabaeus is noteworthy because it ends at the feature once known as 'Dawes Bay,' which marks the line of zero longitude of Mars. (There is nothing deeply meaning-ful in this, simply that for solid planets like Earth, a line of zero longitude has to be put somewhere!)

The far north of this region is marked by many bright ochre deserts – Dioscuria, Cydonia, Eden and Moab. As we move southwards we come to Sinus Sabeus. When Mars is reasonably large, even small- to medium-sized telescopes will show it as a dark curving feature stretching westwards from Syrtis Major and down to the equator, ending in the 'two-pronged' feature called Dawes Bay. Within this region we also find the large crater Schiaparelli. Immediately to the east of Dawes Bay lies Edom. When clouds collect over Edom the region can become rather bright.

Moving further south we come to the bright landmass known as Noachis. Like Argyre (to the west), Noachis can become bright when clouds form over the region.

Polar Caps

Both the northern and southern polar regions of Mars are covered by a permanent ice caps, and both caps can be striking in even a 76 mm telescope. The northern polar region is comprised of a region called Planum Boreum (which means the northern plane). The cap here is comprised largely of water ice and frozen carbon

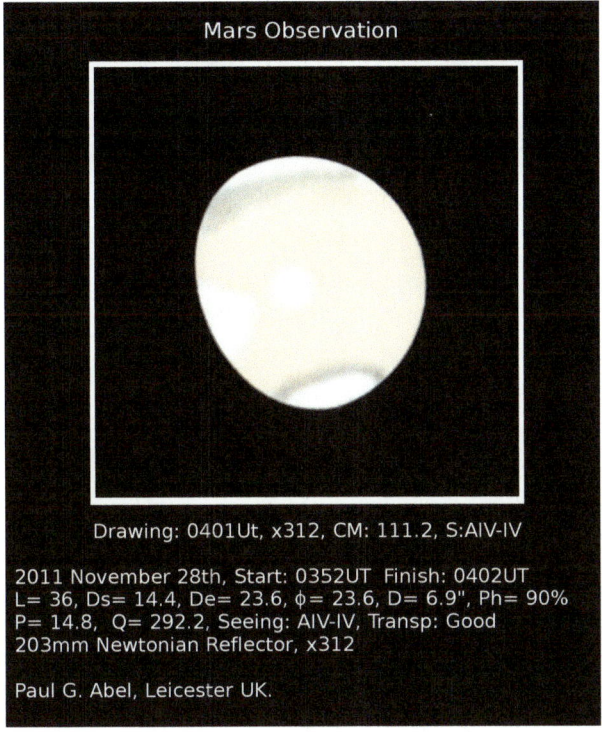

Fig. 5.8 A drawing of Mars made by the author on 2011, November 28, with a 203 mm Newtonian. The northern polar cap is very large and striking

dioxide. As we progress towards the center of the pole we find that the carbon dioxide content increases. Images from spacecraft have revealed dunes and spiral windswept regions very similar to those geometric forms carved by nature here on Earth. The northern polar region is a dynamic place, and avalanches have been recorded here by spacecraft. Perhaps the most impressive feature in the region is the large fissure Chasma Boreale, which is nearly 100 km wide.

The best time to view the north polar cap is during aphelic oppositions. At such times the northern hemisphere is well presented (Fig. 5.8). It is fascinating to watch the brilliant north polar cap shrink and fragment as the northern hemisphere passes from winter into spring. As the temperature rises, the water and carbon dioxide sublimates and returns to the tenuous Martian atmosphere. This causes clouds to form and collect around the volcanoes.

The NASA's spacecraft, the Phoenix lander, touched down on the Martian north pole May 25, 2008, and has since returned a wealth of information about the climate and surface chemistry at the north pole. Among other things, the lander recorded snow falling from cirrus clouds before the probe was buried under vast amounts of ice as winter approached. It took a splendid panoramic view of its environs.

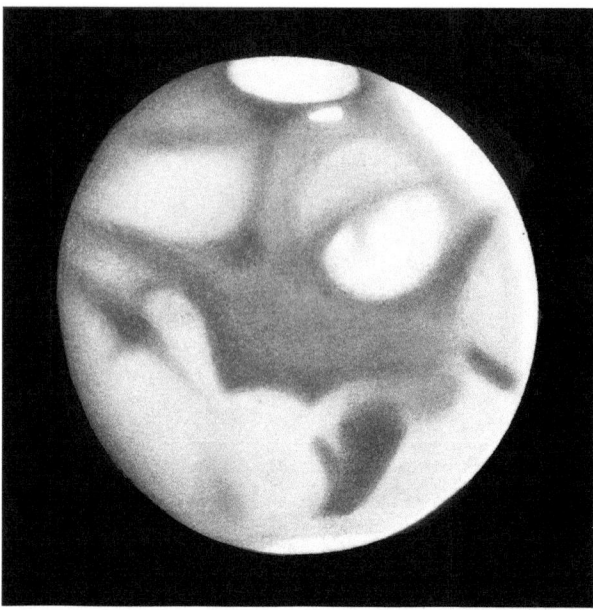

Fig. 5.9 Drawing of Mars made by Dr. Richard McKim, 1988 August 17 01 h 05 m UT, 360 mm OG, ×450, Arcetri Observatory, Florence, Italy (Courtesy Richard McKim)

The southern polar cap is best seen during perihelic oppositions, and this cap grows to be much larger than the north cap and recedes much more quickly during the spring and summer of the southern hemisphere. The reason for this is due to the nature of Mars' orbit. When perihelion occurs (i.e., when Mars is at its closest point to the Sun), it is late spring on Mars, and being closer to the Sun the southern hemisphere (which is pointing sunwards) receives around 45 % more radiation than the north pole does at aphelion (the furthest point from the Sun). As the planet moves closer to the Sun in its orbit, so it moves quicker along it. As a result spring and summer in the south on Mars are shorter and hotter than the ones in the northern hemisphere. At aphelion, Mars is far from the Sun and moves more slowly along its orbit. Thus autumn and winter in the south are colder and longer.

Like the northern cap, the southern polar cap is comprised largely of water ice and carbon dioxide. It is quite remarkable just how large the south cap appears when at its full extent in the winter of the southern hemisphere. As the months progress into spring, the cap retreats very quickly and appears to fragment and splinter. The mountains of Mitchel become visible during the later stages of retreat, and by late spring, the south polar cap appears as a tiny little disk of white, often surrounded by a dark Lowell band. Medium- to large-size telescopes will often reveal that the cap is not uniform and has slightly irregular edges in places (Fig. 5.9).

Historical Observations

The telescopic history of Mars is as fascinating as the spacecraft exploration of the planet, although for entirely different reasons. As we shall shortly see, what we have learned from telescopic observers has provided some interesting insights into the workings of the human mind, as well as the nature of the Red Planet. The exploration of the Red Planet came at a time when new frontiers were being settled here on Earth, and new inventions and technology offered all sorts of exciting opportunities. These events at first glance seem unrelated, but as we shall see, the two are quite intimately connected.

The very first telescopic observers of Mars seemed to have found little of interest. This is not surprising, as their telescopes were rather small cumbersome affairs with narrow fields of view. For most of the time, the Martian disk is small, and it is no wonder that these early optical trunks, as they were called, revealed little about Mars.

The first person to make any real progress with the planet was Christiaan Huygens (1629–1695), a Dutch mathematician and astronomer. Like many learned men of the seventeenth century, Huygens dabbled in everything, and as a result his contributions are many and varied. He discovered Titan, the largest satellite of Saturn. He invented the pendulum clock, as well as making discoveries in calculus and optics.

Huygens was born into a time of great scientific optimism. The heavens were becoming a fashionable playground for scientists to explore the unknown. As one would expect, telescopes were not manufactured in bulk as they are today. If you wanted one, you had to make it yourself. Huygens constructed a 2-in. refractor, and on 1659, 28 November, he pointed it towards Mars, which was a large 17.3″ across. Huygens sketched what he saw. His drawings show a distinctive 'V'-shaped feature; this was of course Syrtis Major (Fig. 5.10). He observed Mars again a few days later, and noted that on December 1, 1659, that Syrtis Major was in about the same position. The only explanation for this could be was that Mars had a 24-h rotational period. With a homemade 2-in. refractor, Huygens had established a correct estimate for the rotational period of Mars.

Huygens' contemporary at the time was the astronomer Giovanni Cassini, who would make many important contributions to astronomy (including discovering the gap in Saturn's rings that bears his name). In 1666, he used the Campani telescope to observe the planet. Alas, it seems he was badly hampered by the poor optics of the time and poor weather. Cassini made some drawings, but it has to be said they did not really contribute anything further.

Robert Hooke in the UK also observed the 1666 apparition. Hooke found he had difficulty seeing the features of Mars in any clear or well defined way, but he was able to make a drawing, and like Huygens', his drawings appear to show the Syrtis Major.

Although the first steps in planetary exploration had been taken, those early astronomers were hampered by the great aerial telescopes of the time. The problem

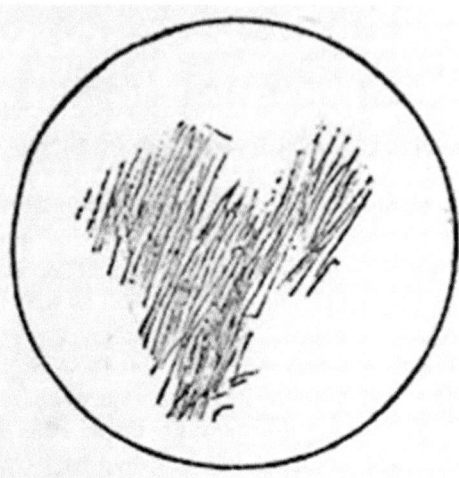

Fig. 5.10 Drawing of Mars made by Christiaan Huygens. The V-shaped Syrtis Major is clearly visible (Courtesy Christiaan Huygens)

was that as the size of the objective lens increased, so, too, did the focal length in order to reduce chromatic aberration. This meant that a telescope of 2 in. diameter could have a focal length of 20 ft! The telescopes were cumbersome and difficult to wield, and these early achievements with these instruments must be admired.

The observers of those early days realized the limitations of their telescopes. Surely there had to be a better type of telescope? The solution came from England.

Isaac Newton needs no introduction. Even those only slightly acquainted with the historical figures of science will know of him. Famous for the first human attempt to formulate the laws of gravity, Newton also dabbled in mathematics, chemistry and astronomy. What many people are often surprised to learn is that he invented the reflecting telescope. In 1672, Newton presented a new type of telescope. It used a concave mirror to collect light, instead of an objective lens. The light was magnified by the primary mirror and reflected up to a flat secondary mirror where it was sent through into the eyepiece. This was a great improvement; not only was the focal length greatly decreased, but so, too, was chromatic aberration.

One astronomer to make great use of this new type of telescope was William Herschel. Herschel was originally from Hanover, but during the war with France, he and his brother came to England. The Herschel family were to make many contributions to science. William would of course discover the seventh planet, Uranus, and make the first attempts at describing the galaxy. His sister Caroline was also an important astronomer in her own right and would discover many comets. William's son, John, would continue in his father's footsteps and would eventually win a gold medal from the Royal Astronomical Society.

By profession, William was a musician, and when he settled in the pretty town of Bath, he made a living composing music and teaching. (His house at 19 New

King Street, Bath, is now the Herschel museum and well worth a visit.) He became interested in astronomy, so, he decided he would make his own reflector, which meant engaging in the laborious task of mirror grinding. In fact, Herschel would become a talented mirror maker, and he would go on to produce telescopes for many of his contemporaries.

Herschel made himself a fine 6-in. reflector and began to chart the sky. On 1781, March 13, he discovered Uranus, and King George III became his patron. Herschel devoted all his time to astronomy. Herschel made a start on Mars in September and October of 1783. The planet was relatively close, and his telescope showed him the south polar cap, which is favorably tilted towards Earth during perihelic oppositions. From his observations he deduced that the true south pole of Mars was not in the center of the south polar cap but rather a little way from the center. Interestingly, he seems to have paid little interest to the dark markings of the planet.

Herschel was a great popularizer of astronomy, and as he published his observations, other astronomers took to observing Mars. In particular, Hershel's discovery of Uranus had a strong influence on Johann Hieronymus Shröter who was galvanized into action by him.

Shröter had trained as a lawyer and was working for the royal court of George III in Hanover when news of the discovery of Uranus reached him. Shröter decided he would dedicate his life to astronomy and resigned his position. He took up residence in Lilienthal near Bremen and took a less demanding job as chief magistrate of Lilienthal. He established an observatory and equipped it with various telescopes made by Herschel. Shröter published all his observations, but it seems his drawings of Mars were lost, and then rediscovered by Francois Terby, who published them after Shröter's death. Shröter observed the features of Mars, but interestingly he believed the dark markings to be little more than cloud formations.

Meanwhile by 1729, optical developments were happening once more. The optician Chester Moor Hall had made a discovery: if a concave lens and a convex lens were put together, then the chromatic aberration could be greatly reduced (since the aberrations of one lens cancel the other). Moreover, the focal length of such a combined lens was much more manageable. The achromatic lens had arrived, heralding the arrival of the large refractors and with it a new type of astronomer: the areographer, astronomers devoted to mapping the surface of Mars.

Johann Heinrich Mädler was born in Berlin in 1794. He showed much academic promise at an early age, but alas he suffered a great tragedy when, at the age of 19, his parents contracted typhus and died, and Mädler was left to look after himself and his younger siblings. The tragedy cut short his plans to go to university, and he settled on teaching instead. He still retained his strong interest in astronomy.

Jakob Beer was by contrast a successful banker who had gone to Mädler for tutoring in advanced mathematics. Mädler seemed to have shared his passion for astronomy with Beer, and shortly after, Beer established an observatory in the Berlin Tiergarten and equipped it with a 3.75 in. (95.2 mm) refractor. From here Beer and Mädler made a great team; they would go on to produce one of the first reasonable lunar maps. Beer and Mädler started on Mars during the opposition of

1830, during which time the planet's southern hemisphere was presented. They became convinced that dark markings on the disk were fixed geological surface features rather than prominent cloud formations like those produced by Jupiter and Saturn, and established a rotational period of Mars of 24 h 37 min 9.9 s. They watched as the south polar cap slowly shrank as the southern hemisphere of the planet passed into summer.

In 1840, Mädler combined all of their Mars observations into a map. Although the map contained many errors, we should still applaud it; after all, it was first ever map of the Red Planet produced by human hand. Other astronomers with larger telescopes continued to observe successive oppositions, and the new larger refractors continued to deliver fine details, so much so that the map drawn by Mädler was rather out of date by the 1860s.

English astronomers started to produce their own maps. Rev. William Rutter Dawes was a clergyman based in Ormskirk. Ill heath forced him to leave the clergy, and like many of his predecessors, he decided he would devote himself to astronomy. (It seems the astronomy bug doesn't just bite – it well and truly savages!) Dawes initially became assistant to George Bishop at his private observatory. Dawes acquired financial independence and set up his own observatory in Kent, then later Buckinghamshire. Dawes was a good observer and an excellent draftsman, and many of his contemporaries considered his drawings to be the best of the time.

The English amateur astronomer Richard Proctor took the drawings made by Dawes and produced a map. The map produced by Proctor had one important difference from that produced by Mädler: names. Mädler had simply labeled the various surface features with letters. Proctor however labeled the features much more evocatively. We have the Kaiser sea, Lockyer land, Dawes Bay and so on. Mädler's map had transformed Mars from a difficult point of light into a planet. Proctor's map transformed Mars from a planet into a place – a place of inland seas, continents and mountains.

Towards the end of the 1800s, new lands and territories were being established on Earth. The same was true for Mars. The United States continued to expand, and the Americans took full advantage of the many good astronomical locations to be found in their country. The U. S. Naval Observatory was established at Foggy Bottom and equipped with a 26 in. (660.4 mm) refractor. From there Asaph Hall would discover the two Martian moons. Large refractor telescopes started to appear all over the United States. Meanwhile, back in Europe the summer of 1877 was the start of something big.

By the summer of 1877, Giovanni Schiaparelli had already established himself as a prominent astronomer. He had explained the origin of the Perseid meteor shower, and his sharp eyesight had allowed him to make many thousands of double star observations. Schiaparelli had studied architectural and hydraulic engineering at the University of Turin, and he graduated in 1854. He wanted to pursue astronomy and decided to embark in some teaching to support his astronomical endeavors. His big break came when he was awarded a small stipend from the Piedmontese government, and this allowed him to complete his training in astronomy in Berlin, then Russia.

Schiaparelli returned to Italy to work at the Brera Observatory, situated on the roof of Brera Palace in Milan. The observatory was equipped with a 8.6 in. (218.4 mm) refractor, and Schiaparelli would use the telescope to make observations there for over 40 years. The opposition of 1877 was a perihelic one, with Mars reaching a good size. Schiaparelli was stunned at the fine details visible on the Martian disk in the telescope at the Brera Observatory.

It sounds counterintuitive, but viewing Mars under excellent conditions is a harder job than observing in ordinary conditions. In moments of good seeing at high powers, Mars appears to be alive with fine contrast details and colors; every part of the surface seems to contain intricate markings and details that threaten to overwhelm you. This seems to have happened to Schiaparelli (although he was color blind, so he would not have seen the subtle pastel shades).

Schiaparelli produced a new map based on his observations. It was revolutionary for two reasons. First Schiaparelli replaced the names of the various features with even more evocative names. The Kaiser sea became Syrtis Major, Lockyer sea became Solis Lacus and so on. Secondly, he introduced the Italian word *canali* to describe the long straight lines he believed he had observed crisscrossing the globe. To the uninitiated, the word *canali* requires little translation. It instantly implies a manufactured waterway (many of which were being constructed in the UK at the time). In Italian it has a more prosaic meaning, and translated means "channel." However the seeds of an idea had been planted. Channels occur naturally, while canals are designed and constructed.

Not everyone was convinced; the English amateur astronomer Nathaniel Green, who was a portrait artist by trade, was not convinced the canals were real. Moreover he disliked Schiaparelli's nomenclature. Green had also produced a map that looked remarkably different. Green concentrated on the whole of the Martian disk, whereas Schiaparelli concentrated more on the fine detail.

Over the coming apparitions, Schiaparelli recorded more details; some of them like Olympus Mons (which he called Nix Olympia) and Libya, were very real. Schiaparelli believed he had observed a new phenomena. Some of the canals seemed to have doubled. He called this bewildering process 'germination'. Schiaparelli had now become convinced of the reality of his canals and began to think seriously about their origin. He noted that rivers on Earth meander around – in short, 'God does not build in straight lines'.

As more Martian apparitions passed, the canals began to gain ground, largely because Schiaparelli was a distinguished and talented observer. An interesting process started to happen. Prominent astronomers, such as Otto von Struve, who had never observed the canals, began to accept their existence, and in doing so added two things – respectability and expectation. If prominent astronomers were seeing them and they were put on the maps, how could anyone else fail to see them?

Sure enough, other observers began to add their own canal observations; Henri Perrotin and Louis Thollon, using the Nice refractor, observed new canals. Perhaps the most powerful avenue of support came from the great Camille Flammarion. During the 1892 perihelic opposition, Flammerion completed his mammoth textbook *The Planète Mars*. The book offered the most up to date information on the

Red Planet, and, needless to say, the canals made an appearance in the book. Although Flammarion had only seen a few canals he accepted Schiaparelli's observations. He gave his own thoughts on the origin of the canals, and carefully dismissed every natural explanation. Flammarion's book was popular, and the reader is left with no option but to conclude the canals are real and artificial.

This was the situation into which the wealthy Bostonian Percival Lowell found himself in when he decided to take up astronomy for the second time in his life. Lowell was born on 1855, March 13, and came from a wealthy family. He went to Harvard and was awarded a distinction in mathematics. After university, he took up diplomatic work in Korea before returning to Boston once more. Lowell had always been fascinated by astronomy, but it seems that it was the reports of canals on Mars by Schiaparelli that excited him and led him back into astronomy.

Lowell decided he would continue the work started by Schiaparelli. He wanted to build his own telescope, and after a long search around the country, settled upon the recently opened territory of Flagstaff, Arizona. Flagstaff was a good choice for two reasons: first it had a post office, and so the railroad passed through the small town, and second its lofty elevation of nearly 7,000 ft means that it gets more than its fair share of good seeing and clear nights. Lowell established his observatory on Mars hill and equipped it with a fine 24 inches (607mm). Clarke refractor. On 1894, May 28, it was set up and ready for action.

Lowell began to observe Mars with the great refractor. Before he had even looked at the planet he was convinced not only that the canals were real but that they were the product of an alien mind. It has to be said, Lowell was a poor draftsman. His drawings of Mars show fixed structures. The real features he recorded are drawn in a rather stark way, and the whole disk was girdled by canals.

Lowell didn't refrain from speculating about the nature of the Martians whom he believed had constructed the canals. He believed the Martians to be a race of brilliant engineers; they had constructed the canals to transport water from the polar ice caps to their cities situated in the vast deserts. Lowell also believed the Martians would be benign; being so advanced they would have put war behind them long ago. When compared to these giants, Lowell found the human race somewhat wanting. He made his views public by often giving lectures about Mars and its supposed inhabitants to packed theaters. He was the showman scientist of his day and wrote many books and articles about Mars, becoming (in the eye of the public at least) the world's leading authority on the Red Planet. His theory that Mars was populated by a race caught in a desperate struggle for survival captured the public imagination.

However, doubts were beginning to surface in the astronomical community. The first such voices commented upon the nature of the planet itself. It was assumed the light areas were land, while the darker features on the planet were seas of liquid water. However William Henry Pickering showed that the light coming from the dark regions was not polarized; if it was water it should be. Further information about the makeup of the planet was coming from the emerging science of spectroscopy. Spectroscopy takes the light from stars and planets and breaks it up, allowing one to see the chemical components of the object emitting the light. W. W. Campbell at the Lick Observatory showed that the spectrum of Mars contained no water.

Meanwhile, in 1887, Edward Emerson Barnard had joined the staff at the Lick Observatory. Barnard had come from a poor background with his education disrupted by the civil war. He initially started work at a photograph gallery but became interested in astronomy. He eventually completed a fellowship in astronomy at Vanderbilt University. E. S. Holder, director of the Lick Observatory, hired him and put him on the observatory staff list.

Barnard was not only an excellent observer but also an excellent draftsman, and in 1894 he used the 36 in. (914.4 mm) telescope at the Lick Observatory to examine Mars. In the fine clear evenings at Mount Hamilton, Barnard found the Martian disk to be full of fine detail and structure. However the canals did not appear to be straight lines; rather, they appeared as a series of broken lines and points and looked to Barnard to be completely natural. Barnard became convinced that the canals were an illusion caused by the eye filling in details of a series of fine broken lines and points.

Lowell did not share Barnard's beliefs; however, Lowell was about to damage his reputation. After observing Venus for a short time, he believed he could see a canal network there. Rather than wait, he published his results, asserting the reality of these structures. This time his contemporaries were not silent, and Lowell received a great deal of criticism. More trouble was in store: experiments had been conducted with globes covered with fine irregular detail. Up close the fine irregular details were obvious. However, when viewed from a larger distance they transformed into a geometric network of straight lines. The implications were obvious. Schiaparelli had seen the fine details of Mars and his visual system had translated them into straight regular lines. Lowell, infatuated with the idea of advanced life on Mars, had happily deceived himself for his good cause. The noted biologist Alfred Russell Wallace, co-discoverer of the theory of evolution, also criticized Lowell. In response to Lowell's book *Mars and Its Canals*, Wallace noted that the thin Martian atmosphere could not really support liquid water and suggested the ice caps at the poles might be composed of carbon dioxide. He concluded that "only a race of madmen would build canals under such conditions."

Barnard was not alone. His European contemporary was Eugene Antoniadi, a Greek astronomer born in Asia. Like Barnard, Antoniadi was a very skilled observer, and his drawings of the Moon and planets were the best of his generation. Antoniadi would become an expert on Mars; indeed he was the first director of the Mars section of the British Astronomical Association.

Antoniadi had been a supporter of the canals. However when he observed Mars during the 1909 opposition using the great 33 in. (838.2 mm) refractor at Meudon he had the same experience as Barnard. In good seeing and at high powers, the straight lines of the canal network were transformed into an irregular collection of fine points and lighter spaces. Antoniadi was unable to see the canal network. Other observers using smaller telescopes began to agree, and Antoniadi published the first Mars map for the BAA that was not covered with canals. Lowell never accepted any of it; he died on 1916, November 12, believing firmly in his theory that the canals were real and the last great efforts of a dying race.

In a sense of course Lowell was correct; the canals were indeed the product of an intelligent culture. Sadly, that intelligence was very human, shaped by a culture that had, for many thousands of years, been putting legendary beings out among the stars. It is easy for us now in the era of spacecraft exploration to see the canal controversy for what it was. However we should recognize that the canal controversy didn't happen just because of poor visual observations. It happened because the observations and conclusions of one man (Schiaparelli) resonated with the perceptions at the time. In Schiaparelli's era, it was largely believed all the planets to be populated. The great astronomers of the day supported the canals whether they had seen them or not, and the great publicists of the day used them for their own ends (however benign).

On a final note, it should be added that, although the canal controversy is a substantial part of the historical record, much good work has been done by amateurs since then. Martian dust storms were recorded in great detail, and the surface changes caused by the storms and the sometimes turbulent atmosphere of Mars were recorded faithfully and in the true spirit of science. We owe these early observers much.

Observing Mars

Mars is a somewhat difficult object to observe. As we have seen, the two different types of opposition bring about different problems. At perihelic oppositions the apparent disk size can reach 25″, but the planet is hopelessly low down in the south for northern hemisphere observers. At aphelic oppositions the planet is high in the sky, but the disk size may not get much beyond 14″. The oppositions either side of these tend to be the best ones for northern hemisphere observers.

It is also true that Mars only really lends itself to close up study with medium to large telescopes. A small telescope (up to 150 mm, say) is unlikely to show much detail until a month or so before opposition (and then for about a month afterward). If you have access to a telescope larger than 150 mm, this will greatly lengthen the time you can spend watching the planet. It is best to start your observations of Mars when the planet reaches a size of 6″. This means you can follow the planet for many months of the year. Obviously we want to follow the planet for as long as possible, as Mars undergoes many seasonal changes that need to be tracked and recorded.

Mars tolerates magnification rather well, and you should be able to use a power of 400× with a 200 mm Newtonian. Indeed during the 2011–2012 opposition, there were some excellent seeing conditions and I was able to use a power of ×500. Of course such conditions are rare, and in general powers between ×250 and ×320 are adequate for serious visual work.

Before going out to the telescope, it is worth consulting *WINJUPOS* to find out what features ought to be present on the Martian disk. However it is worth interjecting a note of warning here. There is a real difference between expecting to see something and seeing something expected! This means we must be constantly careful that we are not observing and drawing exactly what we expect to see. Every time we observe a planet, we must put aside all preconceptions, and observe and record as honestly and objectively as possible.

One thing we should note is the effect that the rotational period of Mars has on Earth-based astronomers. Mars rotates once in 24 h 37 min, so if Syrtis Major is on the central meridian tonight at say 2000UT, then due to Mars' rotation, Syrtis Major will be on the central meridian at 2037UT the next day. This means it takes about 3 weeks to be able to see the whole of the Martian surface from any one point on Earth. This is why cooperation with astronomers all over the world is important during a Mars apparition. Observations from different countries will allow us to monitor the whole surface in a few days, which we could not do from one fixed location. (This is especially helpful if you find the dull face of Mars is presented and you are then inundated with 3 weeks of cloud, after which the dull face has returned to view once more!)

Preparing a Mars Blank

Due to Mars' axial tilt and orbit, there are a number of factors we must take into account when preparing a Mars blank on which to record the observed surface features. We start with a blank circle 50mm in diameter (see the Appendix of this book). Before you go out to the telescope put in the date and the time you plan to go out observing. From *WINJUPOS* (or the BAA *Handbook*) you need to find the following quantities:

- The position angle of the north pole, *P* (in *WINJUPOS* this is the quantity called 'Position angle'). We measure this angle anticlockwise from the central meridian line.
- The maximum angle in defect of illumination, *Q* (in *WINJUPOS* this is called 'Maximum phase'.) Essentially this is the point on the terminator furthest from the center of the disk and closest to the limb.
- Phase. This tells us what fraction of the planet is illuminated. So, if *WINJUPOS* gives the value 0.11 of diameter, this means we have a phase of 89 % (i.e., 100−11=89).

In Fig. 5.11 we construct the Mars blank for the date 2012, June 18, at 2230UT. On this day we have the values P=23.1°, Q=90.1°, and the Phase=89 %. We now need to incorporate this information into the blank. We do this as follows (with north at the bottom):

1. Put a ruler over the disk so it is joining the top and bottom of the disk. If we were to join a line between these points it would the central meridian. Put a small mark at the bottom of the disk (Fig. 5.11a); this is the north point of the disk.
2. Now we put in the position of the north pole. For this date and time, P=23.1°, so using a protractor we measure 23.1° counterclockwise from the north point. This will be the location of the north polar cap (Fig. 5.11b).
3. We now need to put in the terminator. The widest part of illumination is given by the angle *Q*. On this date Q=90.1°, so we measure 90.1° counterclockwise from the north pole. We mark this point on the disk (Fig. 5.11c)
4. Finally we draw in the phase and shade out the part of the disk that is in darkness. The blank is now ready to take to the telescope (Fig. 5.11d).

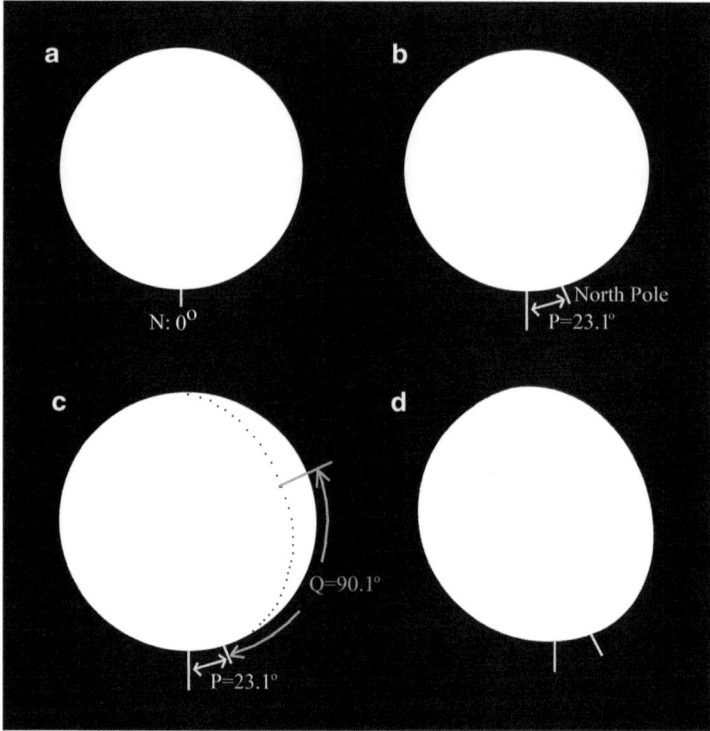

Fig. 5.11 Preparing a Mars blank for 2012, June 18, at 2230UT. Here P=23.1°, Q=90.1°, Phase=89 %. (**a**). The north point shown at 0°. (**b**). The north polar region put in (the angle *P*). (**c**). The maximum defect of illumination (i.e., the part of the terminator closest to the limb) determined from the angle *Q* is put in. (**d**). We shade out the part of the disk outside the terminator to give the Mars blank at the correct orientation. This is the blank we use to draw on

When you go to make a neat drawing for your log book, the above procedure should be used once more to ensure that a correctly orientated Mars blank is used for the final records.

Advanced Projects

Disk Drawings

These are very important, as they track any changes that might occur during an apparition. Before going out to the telescope, you should prepare a few Mars blanks. If it is going to be clear for a long period, make two or three disk drawings.

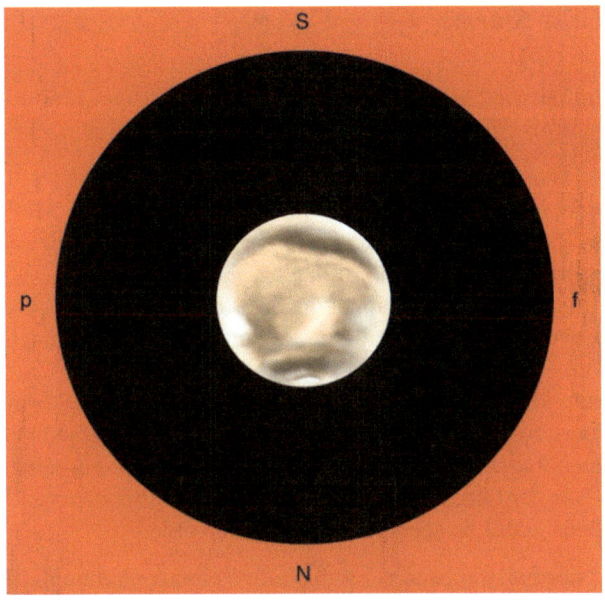

Fig. 5.12 Bright white cloud on the evening limb. Drawing by Carlos Hernandez made on 2012, March 14, 0330UT, with 9-in. Maksutov-Cassegrain telescope, ×310 ω = 209.5 (Courtesy Carlos Hernandez)

Very often morning and evening clouds are present on the limbs of the planet, and these clouds can brighten and change over time (Fig. 5.12). Three disk drawings made over a period of 3 hour will show these changes.

Before you make a disk drawing, make sure you find the right power for the conditions. A power of ×250 is adequate for most purposes when the disk size is greater than 10″. When the disk size is smaller than this, it is best to use a power of ×300 or so. Spend about 15 minutes taking in the view and examining the disk. Once you are familiar with the view, you can begin to make a drawing. Remember that you only have 15 minutes to make the drawing; if you take any longer than this then Mars' rotation will carry the features away from where you have drawn them. It is best to spend the first 7 minutes putting in the most prominent details, such as the polar caps and the dark albedo features. Then note down the time and spend the final few minutes putting in the fine details such as clouds or any regions that can be seen in the albedo features. When you have finished, make sure you include the seeing conditions and magnification used. You should include the CM longitude once you have made your final drawings.

Recording Dust Storms

Despite the tenuousness of the Martian atmosphere, it can produce surprisingly powerful dust storms. There are three types of dust storm that occur on the planet:

- LOCAL DUST STORMS: These take the form of yellow clouds or hazy, obscured regions and are less than 2,000 km in size. They may cover a whole feature such as the Hellas Basin. Local dust storms frequently start in Hellas and Solis Lacus. Local dust storms have also started in Chryse and Argyre.
- REGIONAL DUST STORMS: A regional dust storm takes the form of a bright yellow cloud or clouds that cover a large region (perhaps as much as a whole hemisphere).
- GLOBAL DUST STORMS: These are the most powerful dust storms. When they occur they may cover the entire planet. An example of one such storm occurred when *Mariner 9* arrived at Mars in 1971. The cameras revealed a completely blank disk – the dust had completely permeated the atmosphere. Only the top of Olympus Mons could be seen poking out of the clouds.

Dust storms normally appear as bright yellow clouds. Although small local dust storms are not uncommon in a given apparition, the large global-encircling dust storms only occur when the planet is near perihelion. The dust storm season occurs for $Ls = 180°$ to $330°$ when it is spring and summer in the southern hemisphere and the winds start to pick up. So, during perihelic oppositions, when Ls is between these values, it is best to start the hunt for dust storms.

When recording dust storms, it is just as useful to make a sectional drawing as it is to make a full disk drawing. A sectional drawing will just show the storm and its immediate environment. For example if a small storm has started in Hellas, one would simply draw the Hellas Basin with the storm inside it. It is then easier to compare sectional drawings showing how the storm evolves over time. Dust storms can grow and evolve quickly, so make sure you observe it as frequently as possible so you can record all of the changes taking place.

If you suspect a storm you should report it! A good place to send your observations is the Mars Section of the BAA. The BAA has been studying Martian storms for over a hundred years. Indeed Dr. Richard McKim's BAA memoir, *Telescopic Martian Dust Storms,* is well worth getting. It provides an excellent summary of the many dust storms the planet has produced.

Orographic Clouds

As we have noted earlier, Mars can produce rather bright clouds particularly on the morning and evening limb. Normally cloud activity on Mars is due to the formation and melting of the polar caps. In particular, during aphelic oppositions, when we view the northern hemisphere moves from winter to spring, the northern polar cap melts and returns water and carbon dioxide into the atmosphere, giving rise to the formation of clouds. The clouds collect around the volcanoes of the Tharsis region. They also collect in the vast basins like Hellas, Argyre and Chryse. Indeed Hellas can appear incredibly bright at times when clouds have collected in the basin. Some of the clouds that appear on Mars are at a very high altitude. When these clouds appear on the terminator the sunlight illuminates them and they may appear to extend into the night side of the planet.

Not all bright regions on Mars are due to cloud activity. Some of the bright regions are due to frost that forms on the surface, along with mists and fogs that collect in low lying areas (as they do here on Earth). Like dust storms, clouds are worth watching, as they are dynamic and evolve over a period of a few hours. You should indicate any bright regions you suspect to be clouds in your disk drawings.

Filter Work

Filters can be very useful when observing Mars. The contrast of dust storms and yellow clouds for example are enhanced by a red filter (W#25) or a orange filter (W#21). Orographic clouds are also enhanced by a light blue filter (W#80A).

Another rare phenomena to watch out for is called a blue clearing. Normally, if you examine Mars in a blue filter (say a W#80A or W#38A), the dark albedo features present in integrated light become rather obscure and difficult to see. It may last anywhere from a few hours to a few days. If you see or suspect a blue clearing you should report it. You should also make a drawing of the planet in the blue filter to see if there are any subtle differences in the albedo features when compared with integrated light.

Chapter 6

The Moon

Distance from Earth: 356,410 km (perigee), 406,697 km (apogee)
Diameter: 3,476 km
Synodic Period: 29.53 days
Axial Rotational Period: 27.32 days (synchronous)
Surface Temperature: −184° to +101°
Apparent Diameter: 33′ 31″ (max), 29′ 22″ (min)
Albedo: 0.067 (mean)
Apparent magnitude (brightest): −12.9
Rate of Recession: 3.8 cm/year

A Brief Look at the Moon

There cannot be a celestial object that has caught human imagination more over the millennia than the Moon. Its serene light would have shone down and illuminated the early seas from which all life on Earth sprang. It would have illuminated the night sky for the dinosaurs (presumably they noticed it shining in the sky), and would have provided comfort for our early ancestors as they huddled together in caves, with the Moon banishing the dark and its unseen predators. It is little wonder that the Moon features so much in the writings and musings of human beings.

The Moon is very often the first thing people point their telescopes at. That's reasonable – it's big, bright and impossible to miss. Most people can still remember their first look at the Moon through a telescope. In fact, this is when many are savaged by the astronomy bug. Even a small telescope will transform the lunar disk into a dazzling and bewildering place quite unlike anything seen before. Here is a landscape with vast mountain tops peaking out of the inky black shadows, fault lines and bright rays crisscrossing both ragged landscape and beautiful lava filled planes. There are enormous craters with complex walls carved out by meteorite impacts billions of years ago, when the Solar System was still a very violent and dangerous place. Many must wonder what it must be like to stand in one of the deep lunar craters and watch the Sun rise above it, slowly pushing back the darkness. The lunar landscape is stark, alien and profoundly beautiful. You can never tire of its beauty.

You might think that there is nothing left for the amateur to do when it comes to the Moon. Although the Moon has indeed been studied and even visited by people, there are still a number of useful projects the visual observer can undertake. One thing I should say about this chapter – it is not a guide to the Moon. By that I mean we do not examine the lunar surface in great detail, pointing out various features of interest. There are just so many interesting features, to do them justice would require a book in its own right! So, it is strongly recommended that you get yourself a decent lunar atlas (or a copy of *Patrick Moore on the Moon*) that provides such a narrative. We shall instead take a more general approach; once we have established some useful facts and examined some historical lunar observations, we shall move on to look at the projects that can be undertaken by amateur astronomers.

Orbit and Lunations

The Moon orbits Earth in an almost circular orbit, and the lunar orbit is an example of synchronous rotation. This means that the Moon always keeps the same face turned towards Earth – we never get to see the far side of the lunar surface. As the Moon orbits Earth, the amount of sunlight we see falling on it varies, and as a result, the Moon appears to pass through phases. In Fig. 6.1 is a diagram that shows the lunar orbit and how the phases occur.

We start at Position 1. Effectively, the Moon is between Earth and the Sun and so is not visible. This is called the New Moon. It is at this time that solar eclipses occur (when the Moon passes over the face of the Sun). Of course, a total eclipse does not occur every time there is a new Moon, as the Moon's orbit is not circular and is inclined to the equatorial plane by some 5.9°. If it were (and was a little closer to Earth) there would indeed be solar eclipse each month. This geometry means that not every eclipse is equal; some last longer than others (the longest a solar eclipse can last is 7 min 31 s).

As the Moon passes from new Moon, it reaches Position 2, where it is said to be a waxing crescent. Although the crescent is the only portion of the Moon we can see illuminated by the Sun, it might be possible to see the unlit part (though this is not the far side of the Moon) if Earthshine is present. Earthshine is a phenomenon where the unlit part of the Moon's surface is lit up by sunlight reflecting off Earth.

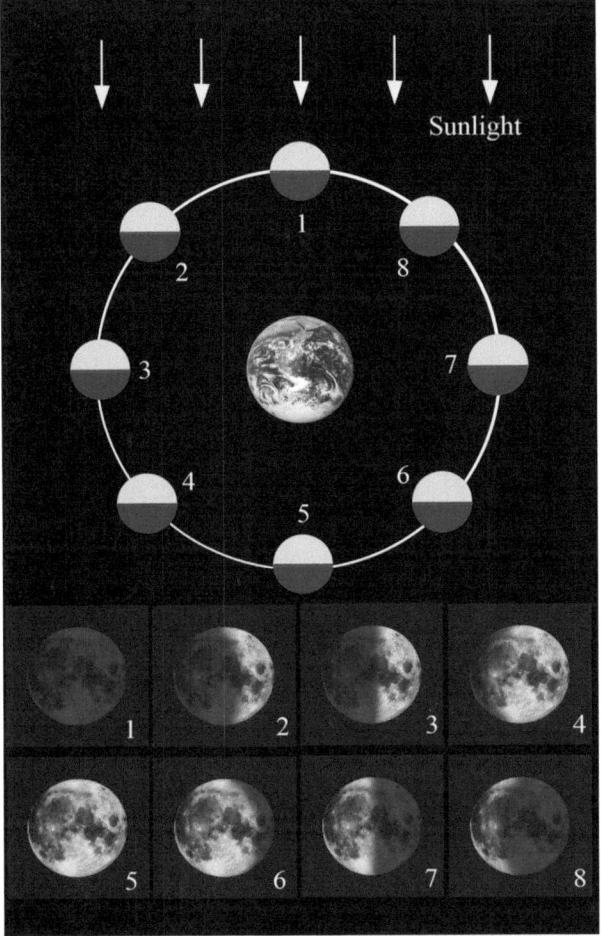

Fig. 6.1 The phases of the Moon as it orbits Earth

It is variable, but sometimes large craters can be illuminated enough to see. Try looking for Tycho Crater when Earthshine is particularly strong.

By the time Position 3 is reached, we have arrived at first quarter. The Moon appears as a half in the sky. This is a good time to begin exploring the Moon if you're new to it. The terminator (the border between night and dark) falls upon some lovely craters, and the southern mountains look particularly fine, their peaks pointing out of the shadows and looking like twinkling stars.

We continue on to Position 4, which is a waxing gibbous phase. Eventually we come to Position 5 – the full Moon. This is the worst time to begin exploring the Moon, as everything is under high illumination, and making out craters and features is practically impossible!

As the Moon moves from full, so it starts to wane; at Position 6 it is a waning gibbous. We then reach Position 7, which is third quarter.

The lunar phase continues to decrease. Eventually Position 8 is reached, and by now it is a waning crescent. The waning crescent usually looks very different from the waxing crescent. When waxing the crescent seems to have more noticeable features (like the Mare Criscium). Eventually the phase decreases completely until a new Moon is reached, and we start the whole cycle again. The period from one new Moon to the next is called a lunation.

Although the Moon only shows us at most 50 % of its surface at any one particular time, we do in fact see a total of about 59 % of the lunar surface. The effect, where objects on the limb of the far-side of the Moon are brought onto the visible limb, is known as libration and is caused by the fact that the Moon's orbit is not entirely circular – it has a slight degree of eccentricity. The zones that can be viewed as a result of libration are called the libration zones. Features within the libration zones are very foreshortened. The effect of libration can be obvious with the naked eye. If you watch the Moon over a period of months, it will be obvious that sometimes the Mare Crisium is well on the disk. At other times it is close to the limb. Programs such as *WINJUPOS* can be used to see which zones have been brought into view by the effect.

The Earth-Moon System

The Earth-Moon system is unique in the Solar System. The sizes of the satellites of all of those planets which have moons is very small compared to Earth-Moon system. With the Earth-Moon system, the Moon is a significant fraction of Earth's mass.

A passing glance through a pair of binoculars reveals the lunar surface to be heavily cratered. This is an historical record from a very violent period in the Solar System's history known as the late heavy bombardment. During this time the Moon, along with Mercury, Venus, Earth and Mars experienced a lot of collisions with large objects.

Figure 6.2 is a cutaway showing the interior of the Moon according to our present understanding. At the very top we have the lunar regolith, or surface. Naturally much of our information about the regolith comes from the Apollo missions. Along with craters and mountain ranges, the surface of the Moon appears to be covered with a fine dust, on top of which sit various larger rocks and rubble. It is thought that constant impacts with micro meteorites has broken the surface up into the fine powder encountered by the Apollo astronauts. (Recall that the Moon has virtually no atmosphere, so meteorites would not burn up as they do here on Earth).

The regolith sits on a crust, and beneath this we have the mantle. Interestingly, 'Moon quakes' have been found to occur deep within the mantle. There is reasonable evidence to substantiate the idea of a partially molten core, and the general consensus seems to be that this core is composed largely of metallic iron and various alloys.

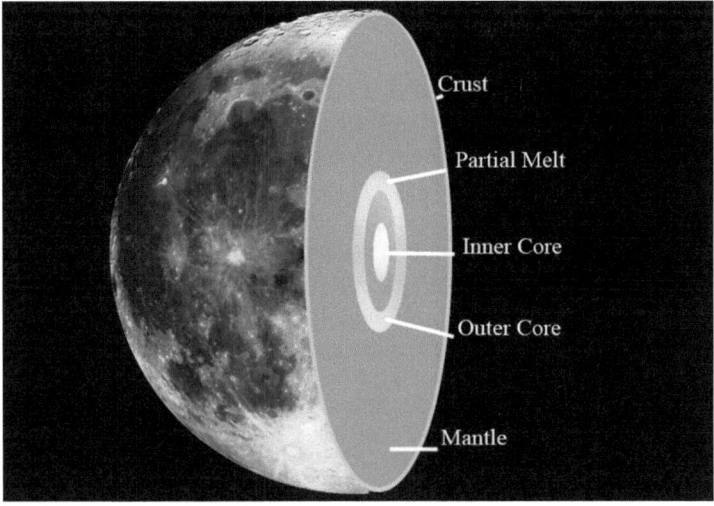

Fig. 6.2 Cutaway showing the Moon's interior according to current models

Lunar Ice

In recent times, space probes have suggested that there might be a small quantity of water ice on the lunar surface. In theory, ice could have been deposited on the lunar surface by comets in the distant past. It was on 2009, September 24, the prestigious *Science* magazine reported that the Indian spacecraft *Chandrayaan 1* detected evidence of water ice.

The LCROSS mission (Lunar Crater Observation and Sensing Satellite) probe was designed to look for water ice at the lunar south pole. At 1131UT on 2009, October 9, the Centaur upper stage of LCROSS' Atlas rocket impacted with 98 km wide crater Cabeus. The resulting impact sent a plume high above the lunar surface which LCROSS passed through, and LCROSS was able to detect the presence of water, albeit in small amounts.

The presence of water ice is quite a remarkable discovery and has immediate implications for a lunar base. Water is of course essential for human consumption, but it can also be used for fuel for rockets. Perhaps, if the will is there, in the not too distant future, we will see lunar bases harvesting the lunar ice for fuel to power spacecraft to the distant worlds of the outer Solar System.

The Origin of the Moon

Perhaps the oldest mystery associated with the Moon is Where on Earth did the damned thing came from! There are a number of schools of thought on this issue, but the theories that still endure are the following:

Capture Theory

Harold Urey, who won the Noble Prize for Chemistry, suggested that the Moon may have formed as an independent planet that passed close to Earth and became captured by Earth's gravitational field. (Variations on this idea include the Moon having formed elsewhere in the galaxy.) However, this theory is very hard to accept since this kind of gravitational capture requires a almost unique set of circumstances. Indeed, most such interactions normally lead to a collision or a change in trajectory, so unless Earth had a vast extensive atmosphere in the distant past when the capture happened, it is hard to see how this model could work. Moreover, the theory does not explain the almost identical isotopic ratios of oxygen that both Earth and the Moon seem to share.

Accretion

In this theory it is suggested that the Earth-Moon system formed together as a binary planet from the accretion disk surrounding the early Sun. Alas such theories do not account for the fact that Earth's iron core is some 50 % of its total mass, while the Moon's is much lower and represents only about 25 % of the total mass. The angular-momentum of the Earth-Moon system is also not explained by this model.

Giant Impact Hypothesis

According to this model, around 4.5 billion years ago, a Mars-sized body collided with Earth. The resulting impact sent large amounts of material out into space, whereby it formed a debris ring surrounding the early Earth. It is from this debris ring that the Moon formed. This model does answer a number of questions. It would explain why, for example, the Moon's core is smaller than Earth's. We know that during the formation of the Solar System there were many planetesimals and that collisions were not uncommon. However there still resides a number of mysteries. The collision with a Mars-sized object would have produced a tremendous amount of heat – indeed theoretical models of such collisions show that the entire surface would have been reduced to a sort of liquid magma ocean. There is (at the time of writing) no evidence that this occurred. Mysteries notwithstanding, this model is the one currently in favor within the astronomical community.

Rather than list interesting features in each quadrant of the Moon (a task which Patrick Moore has done in his superb book *Patrick Moore on the Moon*) we shall include a list of the various types of lunar features that even a small telescope will show on the lunar surface. Figure 6.3 is a map of the Moon showing the general features and craters.

The Maria

These are the large dark regions visible with the naked eye. The word *maria* is Latin for "sea," and they were so named as ancient astronomers believed them to be actual seas. We now know that they are cold ancient lava flows. Dating techniques employed by geologists seem to suggest that the oldest of these lava flows is some 4.2 billion years old. The most recent seems to be the lava flows found in the Oceanus Procellarum – the Ocean of Storms.

It is interesting to note that the maria cover as much as 16 % of the lunar surface, and yet they seem to be located largely on the near side. In the table below is a list of the mare along with their English names and their locations on the Moon. The letter by each of the names is shown on the lunar map in Fig. 6.4.

Impact Craters

Like the maria, the craters of the Moon are probably one of the most immediately obvious and striking features of the lunar surface. Initial attempts to explain the craters of the Moon relied on volcanism. It was largely believed that the craters were volcanic in origin. This is not so outlandish since Earth's surface is furnished with many such fine examples. However it eventually became understood that the craters were the result of impacts on the lunar surface. When a large body such as a comet or asteroid collides with the lunar surface the kinetic energy released in the collision creates a shock-wave starting at the epicenter of the collision, and a large plume of material is thrown out from the point of impact. The end result is a crater and ray material scattered over the lunar surface. The size and nature of the crater depend of course on the size and angle of impact.

Craters on the Moon range enormously in size; some are little more than shallow pits, while others may extend for many miles. A number of them are very deep, and

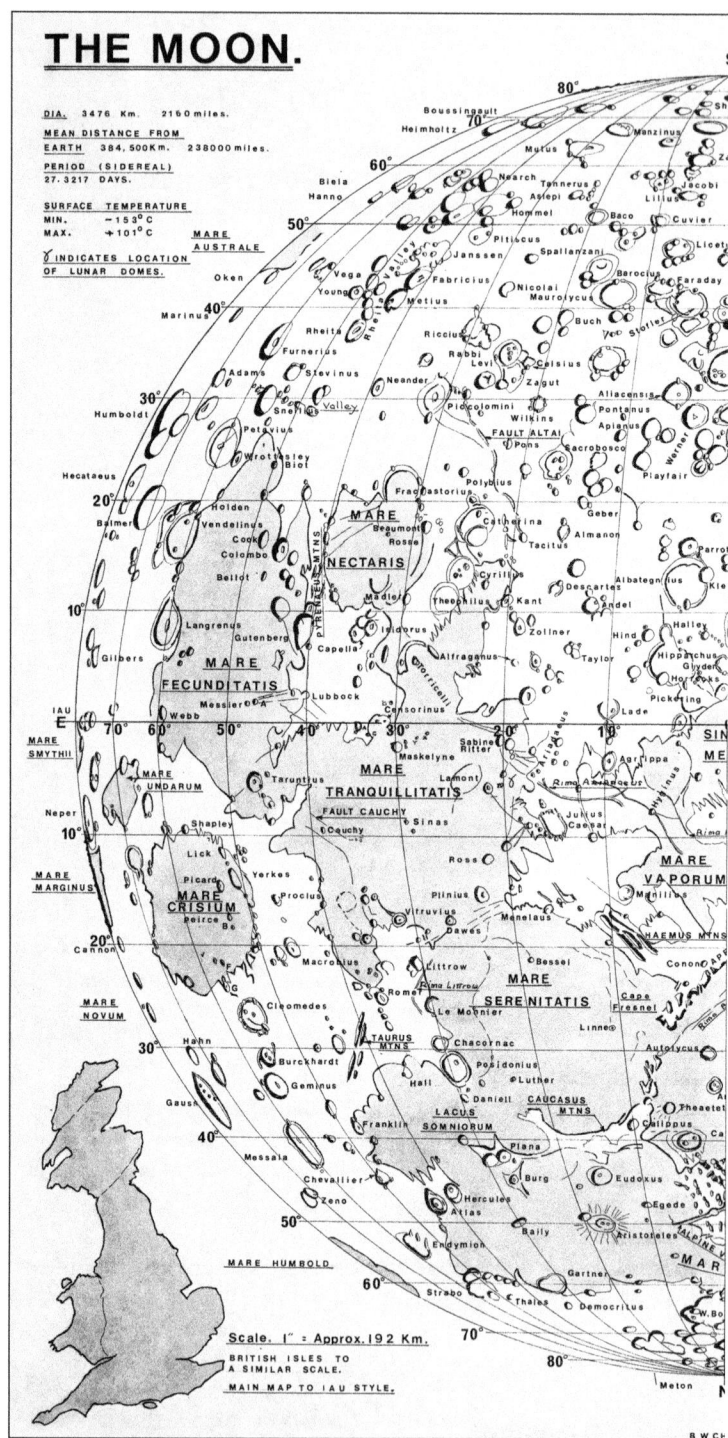

Fig. 6.3 A lunar map drawn by Bert Chapman. This map is still the one used by the

Lunar section of the British astronomical association (Courtesy Bert Chapman)

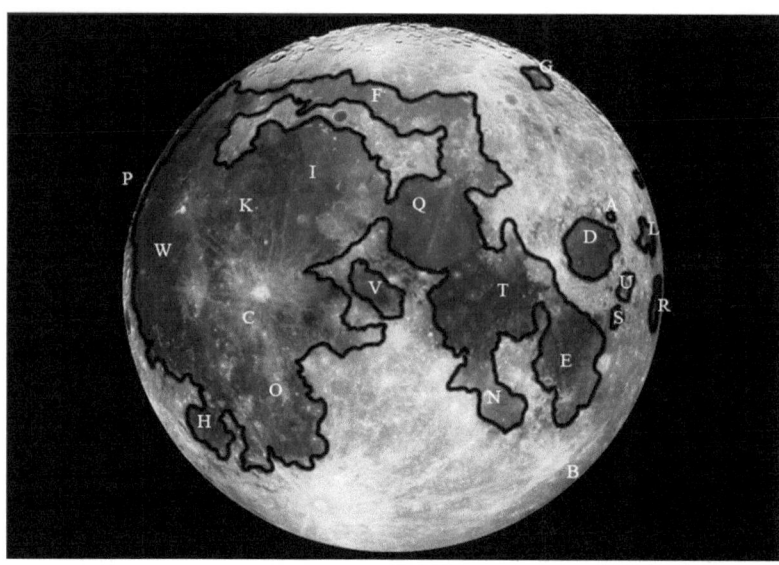

Fig. 6.4 Chart showing the various lunar maria as given in the table below

	Feature	Meaning	Latitude	Longitude
A	Mare Anguis	The Serpent Sea	22.6°N	67.7°E
B	Mare Australe	Southern Sea	38.9°N	93°E
C	Mare Cognitum	The Sea which Became Known	10°S	23.1°W
D	Mare Crisium	Sea of Crises	17°N	59.1°E
E	Mare Fecunditatis	Sea of Fecundity	7.8°S	51.3°E
F	Mare Frigoris	Sea of Cold	56°N	1.4°E
G	Mare Humboldtianum	Sea of Alexander Von Humboldt	56.8°N	81.5°E
H	Mare Humorum	Sea of Moisture	24.4°S	38.6°W
I	Mare Imbrium	Sea of Showers	32.8°N	15.6°W
J	Mare Ingenii[a]	Sea of Cleverness	33.7°N	163.5°E
K	Mare Insularum	Sea of islands	7.5°N	30.9°W
L	Mare Marginis	Sea of Edge	13.3°N	86.1°E
M	Mare Moscoviense[a]	Sea of Moscow	27.3°N	147.9°E
N	Mare Nectaris	Sea of Nectar	15.2°S	35.5°E
O	Mare Nubium	Sea of Clouds	21.3°S	16.6°W
P	Mare Orientale	Eastern Sea	19.4°S	92.8°W
Q	Mare Serenitatis	Sea of Serenity	28°N	17.5°E
R	Mare Smythii	Smyth's Sea	1.3°N	87.5°E
S	Mare Spumans	Foaming Sea	1.1°N	65.1°E
T	Mare Tranquilitatis	Sea of Tranquility	8.5°N	31.4°E
U	Mare Undarum	Sea of Waves	6.8°N	68.4°E
V	Mare Vaporum	Sea of Vapurs	13.3°N	3.6°E
W	Oceanus Procellarum	Ocean of Storms	18.4°N	57.4°W

[a]located on far side

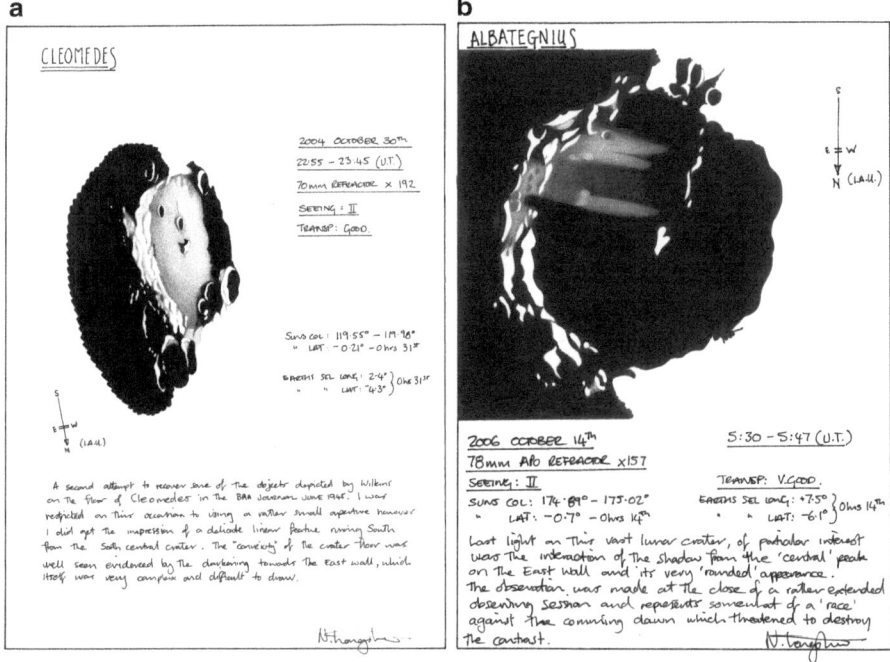

Fig. 6.5 Two superb drawings by Nigel Longshaw showing (**a**) the terraced walls of cleo-medes crater, made on 2004, October 30, between 2255UT-2345UT with a 70 mm refractor ×192 in good seeing, and (**b**) the terraced walls of the fine crater Albategnius. This observation was made on 2006, October 14, between 1730UT-1747UT with a 78 mm refractor, ×157, in good seeing (Courtesy Nigel Longshaw)

the deepest of these is Newton Crater, which has a depth of 6.1 km. It is generally agreed now that the larger lunar craters formed early on in the Solar System's history.

The recent impact craters are very well preserved because there is hardly any erosion on the lunar surface (there being no weather or atmosphere to provide it). In general, impact craters have a sharp outline, and many contain a central peak. Some of the walls of larger craters contain 'terracing.' This is where surrounding material has slumped to form a series of ledges (Fig. 6.5).

The Highlands

The highlands are the lighter regions of the Moon's surface. They are believed to be much older than the darker mare, and, as a result, they are much more heavily cratered.

Domes

These are rounded circular features that are thought to be the remains of shield volcanoes. A number of domes contain a small pit at their summit.

Rilles

These are long, winding, narrow 'channels' that are located normally in the lunar maria (although a few are found in the uplands). They can extend for many hundreds of miles in length and look particularly beautiful when the Sun's angle of illumination fills their interiors with shadow. In general, there are three types of rille:

ARCUATE RILLES	This type of rille looks like a smooth curve and tends to be found near the edges of the mare. It is believed that arcuate rilles form when the ancient lava flows of the nearby mare cooled, shrank and sank into the regolith.
STRAIGHT RILLES	These tend to be long linear features and are believed to be regions of land that lie between two parallel faults that later sunk. The Vallis Alpes provides an impressive example. Another is the splendid Rupes Recta rille.
SINUOUS RILLES	These have the appearance of ancient riverbeds and are believed to be the remains of ancient lava tubes that have subsequently collapsed. They usually begin at the site of an extinct volcano and meander along (Fig. 6.6).

Transient Lunar Phenomena

There is one phenomenon associated with the Moon that is still as controversial today as it has ever been – the so-called transient lunar phenomena (TLP). Essentially, a TLP takes the form of a faint glow (normally dark browns or reds), or obfuscation (like blurring) of crater walls (and sometimes mountains). They tend to be associated with specific lunar features – the crater Aristarchus, for example, seems to be a hot spot, with around a third of reliable observations reporting TLPs seeming to originate in this crater.

The very earliest recording of a TLP comes from twelfth-century Canterbury and involves a spectacular observation by five monks. It was on a summer's evening on 1178, June 18, when the thin crescent of the Moon was visible. The testimony of the monks was recorded in the *Chronicle of Gervase* of Canterbury:

> This year on the Sunday before the Feast of Saint John the Baptist, after sunset when the moon was first seen, a marvellous sign was seen by five or more men sitting facing it. Now, there was a clear new moon, as was usual at that phase, its horns extended to the east; and behold suddenly the upper horn was divided in two. Out of the middle of its division a

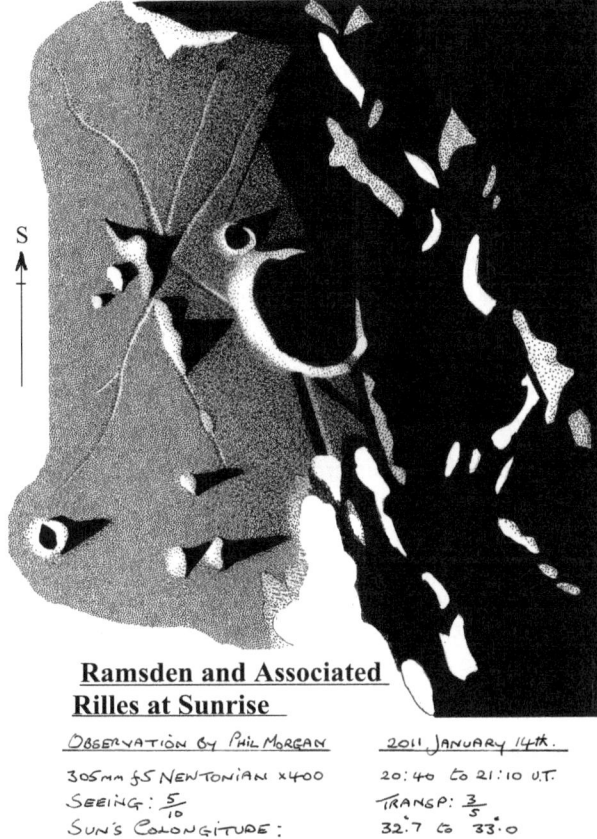

Ramsden and Associated Rilles at Sunrise

OBSERVATION BY PHIL MORGAN   2011 JANUARY 14th.

305mm f5 NEWTONIAN x400   20:40 to 21:10 U.T.

SEEING: $\frac{5}{10}$ TRANSP: $\frac{3}{5}$

SUN'S COLONGITUDE: 32.7 to 33.0

Fig. 6.6 Ramsden and associated rilles at sunrise, drawn by Phillip Morgan on 2011, January 14, 2040UT-2110UT with a 305 mm Newtonian reflector ×400, average seeing (Courtesy Phillip Morgan)

burning torch sprang, throwing out a long way, flames, coals and sparks. As well, the moon's body which was lower, twisted as though anxious, and in the words of those who told me and had seen it with their own eyes, the moon palpitated like a pummelled snake. After this it returned to its proper state. This vicissitude repeated itself a dozen times or more, namely that the fire took on tormented forms variously at random, and afterwards returned to its prior state. Even after these vicissitudes, from horn to horn, that means along its length, it became semi-black. This to me who writes this was told by those men who with their own eyes saw it, and who are willing to swear an oath that they have not added to nor falsified the above written.

It has been suggested (by Jack Hartung) that what the monks of Canterbury observed was a collision with the Moon and a comet or an asteroid that resulted in the creation of crater Giordano Bruno. Since then there have been a number of reliable reports by both amateur and professional astronomers. Some of the most pertinent reports include:

- 1787, APRIL 19–20: Sir William Herschel observed three red glowing spots on the unlit part of the Moon, a phenomena that he tried to explain (albeit incorrectly but, understandably) as the result of active volcanism.
- 1958, NOVEMBER 3: Niklai A. Kozyrev using a 48 in. reflector at the Crimean Astrophysical Observatory observed what appeared to be a bright light at the central peak of crater Alphonsus. More importantly, he managed to take a spectrum of the event, which he claimed showed evidence that a bright gaseous emission had taken place, with the molecules C_2 and C_3 appearing in the spectra.
- 1963, OCTOBER 29–30: James A. Greenacre and Edward Barr were observing the Moon with the Lowell telescope and recorded bright reddish, pink and orange glows in both the SW corner of Artistarchus and at Cobra Head.
- 1992, DECEMBER 30: Audouin Dollfus, using the Paris observatory, recorded unusually bright regions in Langrenus Crater.

Since then, there have been numerous other reports from reliable observers. What does it all mean? It is quite clear that the Moon is a dead world with no active volcanism. Indeed there has been no geological activity on the Moon for billions of years. There are a number of possible explanations. First it could be that TLPs are a phenomena of Earth's atmosphere rather than actual lunar phenomena. It is certainly the case that poor seeing creates all sorts of distortions; false color and blurring of craters occurs quite readily when the seeing is less than ideal.

If the phenomena are real, a number of theories have been suggested. First, TLPs could be the result of 'outgassing' on the Moon. The theory goes that occasionally, small caverns underneath the surface release pockets of gas that well upwards, causing the glows and clouds in the regions in question. Indeed, results from the Lunar Prospector spacecraft support the idea that there has been outgassing of radon to the surface. Yet another theory suggests that at sunrise, the dust on the lunar surface rises and swirls about. Occasional meteorite impacts would also create glows and clouds if the impacts were large enough.

It seems most amateur astronomers have an opinion ranging from 'They don't exist at all' to complete conviction in their reality. Of course the only scientific approach to take is to make observations and collect data. The matter will probably continue to rumble on until such times that multiple clear images, drawings and indeed spectra of the same TLP event are finally, if ever, recorded.

Historical Observations

As you might imagine, the historical account of lunar observations is a rich seam. Indeed, to fully cover the relationship between humans and the Moon would require a book in its own right. I am, therefore, only going to paint a general picture. A more fuller account can be found in the excellent book *Epic Moon* by William Sheehan and Thomas Dobbins.

The first recorded map of the Moon was made by William Gilbert sometime around 1600. This map was based on Gilbert's observations made with the naked eye and show quite distinctly the dark mare regions that had become thought of as seas. The map was not published until 1651, but by then the instrument that was to revolutionize astronomy – the telescope – had been invented.

It seems that no one person can claim to have invented the telescope. It is thought that its invention occurred at sometime around 1608 in Holland. A number of people tried to file patents, but the simple refracting telescope (known then as optical trunkes) was deemed too simple to patent!

Galileo is credited with being the first telescopic observer to publish a map. This is true, but an English astronomer and philosopher by the name of Thomas Harriot appears to have preceded Galileo in producing a telescopic chart that he never published. His drawings show a number of large craters as well as all of the dark mare.

To be fair, though, Galileo made a much more extensive survey of the Moon, and although his telescope produced a magnification of ×50 or so, he became an avid observer and made some 150 drawings showing various lunar features. Galileo was more than just a fine observer. He also had a deep desire to explain what it was he had observed. He realized that the circular craters he was observing were depressions in the surface. In a delightful example of renaissance ingenuity he observed the lengths of the shadows cast by the lunar mountains, and by using simple trigonometry, he was able to estimate the height of the lunar mountains and found that they were decidedly lofty! So it began, the extension of the physical universe. Galileo had transformed the Moon from a perfect divine sphere residing in the unreachable, unknowable fathoms of heaven into a world, a place of deep depressions and vast mountain ranges that were all illuminated as the Sun passed overhead. Perhaps more than anyone else, Galileo started the humanization of space.

Johannes Hevelius (1611–1687) was a wealthy brewer who lived in Danzing, a settlement located on the Baltic Sea in northern Poland. Hevelius had inherited his father's fortune and was free to do whatever interested him. In a way, he was the father of lunar topography (the study of lunar features). Astronomy became his passion, and at some point in the 1640s he established the Sternberg Observatory, which was equipped with that cumbersome telescope of the era – the aerial telescope. Essentially, in order to overcome the problems of chromatic aberration, the solution at the time was to make the telescopes as long as possible.

The Sternberg Observatory telescope was some 150 ft in length! It was driven by a system of ropes and pulleys. Quite frankly, how these early astronomers managed with these great cumbersome devices is a mystery! The slightest breeze must have made the image dance and shudder in the most alarming manner. It is perhaps no surprise that Hevelius did much of his lunar work with smaller focal length telescopes.

Hevelius produced his own book on the Moon that he privately funded. The text included some 45 copper plate drawings, and of course a lunar map. He had introduced a scheme for naming the various lunar features that have long since been abandoned. Indeed, they were abandoned in Hevelius' time when his nomenclature was challenged by Giovanni Riccioli (1598–1671). Riccioli was a professor of mathematics at Bologna University.

Interestingly, Riccioli wrote an enormously long treatise on both the practical and theoretical applications of the emerging science of astronomy. His rather long and detailed *Almagestrum Novum* also contained a lunar map. The map was compiled from drawings from his assistant Franscesco Grimaldi and Hevelius himself. It is here that we find the first use of the names we still use today. Riccioli gave the dark mare some wonderful enigmatic names – the Sea of Showers, the Sea of Tranquility and so on, and these names stuck. Riccioli also introduced the scheme of naming craters after great scientists, philosophers and celebrities.

Perhaps the next person of interest in the study of the Moon is Tobias Mayer (1723–1762). Mayer was an interesting character and was in many ways a self-taught man. He had no formal training, so he taught himself technical drawing and mathematics. Using a tiny telescope by today's standards (a 35 mm refractor), Mayer used a micrometer to measure specific points on the lunar surface and became the first person to use selenographic coordinates, the latitude and longitudinal systems used to identify the positions of lunar features. He produced a number of maps and started work on a lunar globe.

Mayer also made another important observation. In the summer of 1753 he used the Gottingen observatory to measure the position of the Moon with respect to certain stars. Such measurements would help astronomers refine the Moon's orbit further. Along with noting the inaccurate recordings of positions of certain stars in the modern star charts of the time, he also observed a number of occultations – the passing of the Moon in front of a star. He observed that as the star appeared to approach the lunar limb, the starlight did not slowly fade or flicker – it simply winked out when the lunar surface covered the star and re-appeared after the occultation. If there had been an atmosphere, the starlight would have been refracted as it passed through the lunar atmosphere. By direct observation, Riccioli had shown that the Moon must be an airless world.

As we move into the modern era, particularly the start of the twentieth century, the Moon became the domain of the amateurs and received little attention from professional astronomers. The Lunar Section of the British Astronomical Association was probably the most active group at the time. By then emphasis on lunar observation had become about map making. It was thought that small changes still occurred on the lunar surface, and the TLP were seen as evidence of this. In order to detect these small changes, one needed maps showing as much detail as possible. Thus the objective was to produce maps with as much fine detail as possible, along with detailed maps of the libration features.

Perhaps one of the most avid lunar cartographers (or selenographers, as they were known) was the Welsh astronomer Hugh Percy Wilkins. Wilkins worked for the (now defunct) ministry of supply. He built his own telescope (a 15 in. (381 mm) reflector) in the garden of his house in Bexley Heath and observed the Moon as frequently as possible. Wilkins was later joined by a young Patrick Moore (who was an excellent lunar draughtsman), and the two would frequently go across to the

Meudon refractor and use its splendid 33 in. (838.2 mm) refractor to search for finer and finer details.

The end result of this was what Wilkins regarded as his great life work – a large 300 in. lunar map. Although Wilkins was a great popularizer of astronomy and encouraged many young people to take part (including Patrick Moore), he was not a good draughtsman. Moreover he was prone to recording details that were simply not there. As a result, the map is something of a crowded mess, and virtually useless. Wilkins was something of a celebrity astronomer of his day, and this too would get him into trouble.

An example was the famous O'Neill's bridge incident. John O'Neill was the science editor for the *New York Herald Tribune*. In July 1953 he was observing the Moon with his 4 in. (101.6 mm) refractor when he happened upon an unusual looking feature in the Mare Criscium. He noticed a small fan-shaped wedge of light on the floor of the Mare Crisium and believed it to be a natural bridge. The story naturally caused a sensation. The year 1953 was the height of the flying saucer era, and a number of people at that time had already suggested that aliens might have built bases on the Moon.

In August, Wilkins observed the area under the same illumination and apparently confirmed O'Neill's observation (although it turns out that Wilkins had viewed a different feature). Wilkins then got rather caught up in events and gave a radio interview whereby he seemed to suggest that the bridge had "popped up overnight" and was artificial. There was great alarm among his colleagues at the BAA, who were concerned that amateur astronomy might be regarded in the same vein as flying saucers. Wilkins continued to support the idea that the bridge was artificial and made further observations using the 60 in. (1524 mm) telescope at Mount Wilson. This, along with his flirtation with the flying saucer movement, was seen to have damaged Wilkins credibility, and he resigned from the BAA.

It is interesting to note that although Wilkins seemed to be convinced the bridge was artificial, Patrick Moore was not. There is a lunar log book where Patrick observes for himself the so-called bridge and concludes it is a natural formation. The O'Neill bridge affair is rather a sad one, and it may be unfair to chastise Wilkins too harshly from the safe environs of the twenty-first century. It is clearly the case he was somewhat hasty in his judgments, but as we shall see later in the case of Percival Lowell, he was a man of his time, and in a lot of cases, we cannot separate people from the time they lived in, as they are acolytes for the events and ways that define that era. He was a man who was dedicated to astronomy and encouraged many to make a start in the subject.

Before we finish this section we must mention one final person who takes us up to fairly recent times – Harold Hill. Harold Hill was an amateur astronomer based in Wigan, England. He used a variety of telescopes to examine the lunar surface. Hill may not only have been the finest lunar observer of his era but also the finest draughtsman. In fact Hill, like Wilkins, came from an era when recording finer and finer detail was the aim of lunar observing. Hill however focused his efforts on the southern uplands, a region that had not been well covered by lunar spacecraft, and he spent many years observing and drawing the mountains and valleys found in the lunar south pole. Although these regions have now been photographed, Hill's work still stands out as an impressive feat.

The lunar observers of the twentieth century perhaps had the hardest time. Their observing careers spanned both the pre-Space Age and the Space Age. When the first probes examined the Moon and began to send back photographs, it was very soon realized that the amateur contribution to lunar mapping had come to an end. Since this was, for the most part, the main occupation of organizations such as the lunar section of the BAA, it would be understandable if the visual observers at the time had given up hope and abandoned the Moon. Thankfully this did not happen. They continued to show that amateurs can adapt and continue to play an important role, and the lunar section today continues to do fine work. As we shall see later on, there is still plenty of work left to be done on the Moon.

Observing the Moon

The Moon is big and bright, but that does not necessarily make it easy to observe! A 3 in. (76.2 mm) refractor is enough to undertake serious study of the Moon, and a telescope this size will reveal many different craters plus the beautiful southern uplands. As with observing the planets, it is important to get the magnification right; if the seeing is poor then don't use high powers!

Drawing the Moon requires a great deal of care and patience; it is not overly difficult provided you proceed in a logical fashion. By far the most common type of mistake is to draw too wide an area of the lunar surface. Often the beginner is tempted to draw a vast swathe of land surrounding a crater or some other object of interest. What you will find is, as you start drawing a large area, there is so much detail, to record it all could take many, many hours! Restrict yourself to one small feature.

It is a good idea to draw your chosen lunar feature as large as possible. Even with a small feature, a lunar drawing may take an hour or more to complete. The first thing to do is fill in the outline of the feature, so if it is a crater, begin with the rim and central peak – the obvious points. If the terminator is present, make sure you draw the regions closest to the terminator first. This is important because the terminator is slowly moving and not static. Indeed you can watch the lofty mountain peaks emerge from crater shadows over the course of an hour!

With the basic features in place, you can then put in the fine details, and use your basic outline to make sure they go into the correct position. Don't be too disheartened if you find your early drawings are a little inaccurate with regards to the positions of features. It does take some practice to get the placing of tiny lunar features correct!

Some people like to make the outline drawing of the crater before they go outside. This goes against the practice of visual observing – to record what we see as honestly and as accurate as possible. You will find lunar drawing is a great way to improve your positional accuracy.

Along with your drawing, you should also include the following information:

1. The name of the feature(s) observed.
2. The date and the time you started and finished the drawing (in UT).
3. Seeing conditions,
4. The Moon's age (how many days from new Moon).
5. Illumination (i.e. the phase in %).
6. The Sun's colongitude at the start and end of the drawing (this tells us where the terminator is located)
7. Your name and location.
8. Telescope and magnifications.

You can get points 4–6 from *WINJUPOS*, which supplies all of this information in its ephemerides in the lunar section.

Advanced Projects

As we have seen, much of the amateur effort in the past has been concerned with lunar cartography and recording detail as fine as possible. Spacecraft images have of course long since superseded these efforts, but this doesn't mean that there is nothing to be done. In fact the Lunar Section of the British Astronomical Association has a number of projects that require amateur participation and can be of real scientific use. Included here are the basic outlines of these projects, but a more comprehensive discussion can be found on the Lunar Section website (http://www.baalunarsection.org.uk/)

Transient Lunar Phenomena Repeat Events

One of the theories attempting to explain TLP events concerns the nature of the craters themselves. It has been suggested that certain parts of specific crater floors might contain material that is of a different color and can be more reflective than the rest of the crater. As a result, under certain angles of illumination, the Sun makes this part of the crater seem brighter and perhaps more colorful than it would ordinarily appear. If any of the previously recorded TLP events (and there are over 700) are the result of this mechanism, then the events themselves should be repeated when the angle of illumination is repeated.

The TLP section of the BAA Lunar Section produces a monthly list of all previous TLP events whose angle of illumination, libration and phase will all be the same as the time of the original TLP report. All you have to do is go out and locate the feature in question and see if there is anything unusual about it. This work is extremely valuable and may well yield results that some of the TLP events of the past are indeed just the effect of brighter material covering parts of some crater floors.

Lunar Occultations

Clearly the Moon subtends a large angle in the sky (around 30″ or half a degree), and so it occasionally passes in front of planets and stars. This event is called a lunar occultation. Sometimes, the star or planet is not completely occulted and simply passes very close to the lunar surface; this is called a graze.

For occultation observations, we simply record the time when the star or planet goes behind the Moon (this does require an accurate clock – you can use a radio clock that updates itself so it is always at the right time). You should also record when the star (or planet) emerges from occultation.

When a graze occurs, the object being occulted may 'wink' on and off as it passes behind the lunar mountains and valleys. If this occurs you should also record these details.

Low Light Studies

There are not many images from spacecraft showing lunar features under low illumination. When the Sun is low down from the point of view of a lunar feature, it will cast long shadows. At such times, the long shadows reveal topological details that might be invisible under higher illumination. Studies of various craters and features at low angles of illumination might well reveal new features previously unrecorded.

High Light Studies

Here the objective is to study various features when they are under direct illumination, with the Sun virtually overhead. There are very few observations of lunar features under high light, since most people observe the Moon when a feature is on or near the terminator, where there is a nice balance between light and shadow. The aim with high light studies is to examine various features to see if any new rays can be found (rays are well emphasized under high light). This is also a good time to examine features to see if there have been any changes in albedo.

Studies of Specific Features

BANDED CRATERS A number of craters have 'dusky' radial bands (Aristarchus is a fine example). Conventional theories tell us these bands were made during the impacts that formed the crater, and that as time passes, space weathering causes these bands to fade, which is why older craters appear not to have them. There is some disagreement with this theory, and the Lunar Section hopes to establish a relationship between crater banding and the size of craters, since there are a number of older large craters that have no banding present at all.

RILLES The aim with this work is to build up a more complete and up to date catalog of the rilles. In particular, the BAA hopes to produce a catalog of rilles that records not only their location but also their size and type (i.e., sinuous, arcuate). This work involves examining rilles and classifying them according to type (Fig. 6.7).

RAYS A similar catalog needs to be made of the many rays that can be found all over the surface of the Moon.

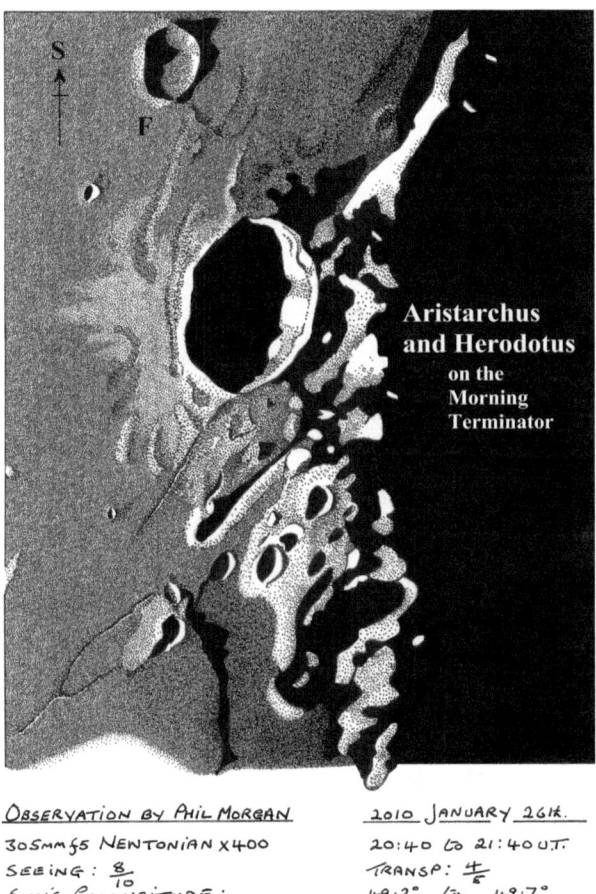

Fig. 6.7 Splendid drawing by Phillip Morgan of crater Aristarchus and Herodotus, made on 2012, January 26, between 2040UT and 2140UT using a 305 mm Newtonian reflector, ×400 in good seeing

Domes A good catalog of lunar domes now exists, but an area of interest associated with them are summit pits. In particular, it is of interest to know whether the summit pits of domes are always in the center of the dome and whether they are associated with the volcanic nature of domes or simply the result of random meteorite impacts.

Many of the projects suggested here can be carried out with a modest telescope. Certainly a telescope of 4 in. (101.6 mm) aperture should be sufficient. Moreover, if you live in a heavily light polluted city, the bright light of the Moon will not be affected. Although our closest neighbor has been extensively studied and even visited, there are still a few mysteries left to solve, and any telescope time you assign to Earth's ancient companion will not be in vain.

Chapter 7

Jupiter

Useful Facts

Distance from the Sun: 815,700,000 km (max), 740,900,000 km (min)
Diameter: 142,884 km
Sidereal Period: 11.86 years
Synodic Period: 398.88 days
Rotational Period: System I (Equator): 9 h 50 min 30.003 s
System II (Everywhere Else): 9 h 55 min 40.623 s
System III (Magnetosphere): 9 h 55 min 29.711 s
Mean Surface Temperature: –150 °C
Apparent Diameter: 50.1″ (max.), 30.4″ (min.)
Albedo: 0.43
Brightest Apparent Magnitude: –2.6
Satellites: 66 (known)

A Brief Look at Jupiter

Jupiter is a very alien world. All we can ever see of it is its colorful tumultuous atmosphere. It is a planet of Earth-sized storms, powerful lightening and almost surreal meteorology. The multiple clouds bubble up from the deep and interact with the storms in a chaotic, bewildering dance. Of all the worlds in our Solar System it is the most accessible. Even the smallest telescopes will reveal its cloud belts and the four large Galilean moons.

P.G. Abel, *Visual Lunar and Planetary Astronomy*, The Patrick Moore
Practical Astronomy Series, DOI 10.1007/978-1-4614-7019-9_7,
© Springer Science+Business Media New York 2013

A view through a 4 in. (101.6 mm) telescope will reveal the two main dark equatorial belts and the Great Red Spot when it is prominent. Apertures of greater than 8 in. (203 mm) reveal a myriad of details, subtle shadings and oval storms, all in a variety of pastel tones. There can be so much to see in fact that in moments of good seeing, it is an enormous task to record all of the details in time before the planet's swift rotation carries them out of sight.

Jupiter has long been the domain of the amateur, and in particular, the Jupiter section of the British Astronomical Association has a proud history of observations and analysis spanning over a hundred years. Today we have the splendid results of numerous space missions, and our understanding of this giant planet has greatly improved in the last few decades.

This does not mean that the work for the amateur is complete – far from it. We can use our telescope time to provide a long series of observations. By providing observations over many apparitions we can monitor global changes, not just localized ones as revealed by spacecraft, and we can begin to understand the long-term behavior of the planet.

The Jupiter section of the BAA still has many hundreds of active observers contributing their observations. Those results, together with spacecraft data, form the basis of an ongoing narrative of the planet. The planet is constantly evolving, and Jupiter provides the amateur astronomer with an enormous number of possible contributions and potentially exciting discoveries.

Orbit and Apparitions

Jupiter orbits the Sun approximately once every 12 years, and so it spends about a year in each of the constellations of the zodiac. (The zodiacal constellations are simply the ones that contain the ecliptic – the Sun's path across the sky. It is also the orbital plane of the Solar System for the planets, and so, unsurprisingly, the planets travel along, or very close to the ecliptic.)

Sagittarius and Capricorn are far below the celestial equator. When the planet comes to opposition in these constellations (normally in the summer months in the northern hemisphere), it is always low down, and the image of the planet is distorted by Earth's atmosphere. Since the planet is so large (the apparent diameter of Jupiter never falls below 30″), a low to medium power can be used and useful observations can be made.

By the time the planet moves into Aquarius, there is usually a great deal of improvement in image definition. Of course the best times for northern hemisphere observers are to be had when the planet is well north of the celestial equator. When the planet resides in the constellations of Taurus and Gemini opposition occurs in the winter months when the nights are very long. Jupiter rotates once in just under 10 hours, and on the winter solstice, northern latitudes can experience as much as 15 hours of darkness. This means that winter oppositions provide us with the opportunity to view the entire planet in one night!

Jupiter has a very small axial tilt of only around 3.1°, and so for all intents and purposes it experiences no seasons, unlike Earth (and Mars and Saturn), which have appreciable axial tilts and well defined seasons of winter, spring, summer and autumn.

There is another cycle which is of interest. Jupiter's apparent diameter at opposition varies from 46" to an enormous 50" across, and it achieves the 50" every 6 years. The reason for this is simple – it is due to the orbit of the planet around the Sun. Jupiter's orbit is slightly eccentric, and this means that it is ever so slightly closer to the Sun (and therefore to us) at perihelion than it is at aphelion. Since it takes the planet about 12 years to complete one orbit, it therefore takes it half that time (i.e., 6 years) to pass from aphelion (where the disk size is smaller) to perihelion (when the planet is closer to us and therefore presents a larger disk size).

The Composition of Jupiter

The first four inner planets, Mercury Venus, Earth and Mars, are all small rocky planets. Jupiter is a very different world and belongs to a class of planets known as the gas giants. Gas giants have huge, extensive atmospheres, and no solid surface. As a result they undergo what is termed differential rotation. In practice this means that the equator rotates quicker (9 h 50 min 30.003 s) than the rest of the planet (9 h 55 min 40.623 s), and it is this which gives rise to Jupiter's distinctly flattened appearance. Even a cursory glance in a small telescope reveals the planet has an extensive bulge at the equator.

Figure 7.1 is a schematic based on the currently accepted planetary models, showing the proposed interior of the planet. At the top we have a rather extensive cloud deck. It is believed that the cloud deck is comprised of three layers. At the top we have the cold bright clouds composed of ammonia ice crystals. Beneath this we have a denser cloud layer composed of ammonia hydrosulfide or ammonium sulfide. The bottom cloud deck is thought to be composed of water. It is thought that this cloud deck is the most dense of the three and contributes to the impressive dynamics of the upper Jovian atmosphere.

As we pass through the cloud deck we come to a vast region composed of molecular hydrogen and helium. As we journey even further down, the temperatures and pressures become incredibly high; the hydrogen and helium are literally squashed into a very peculiar substance known as 'metallic' hydrogen and helium. It is thought that this vast churning metallic sea is what gives rise to Jupiter's enormous magnetic field. (Indeed, if it were visible Jupiter's magnetic field would appear as large as a full Moon). It is thought that there is a small rocky core at the center of the planet, but for obvious reasons, there is very little data about the conditions of the planet beneath the cloud layer.

Jupiter gives off far more heat than it receives from the Sun. There are two explanations for this extra source of heat. The first is a form of primordial heat, left

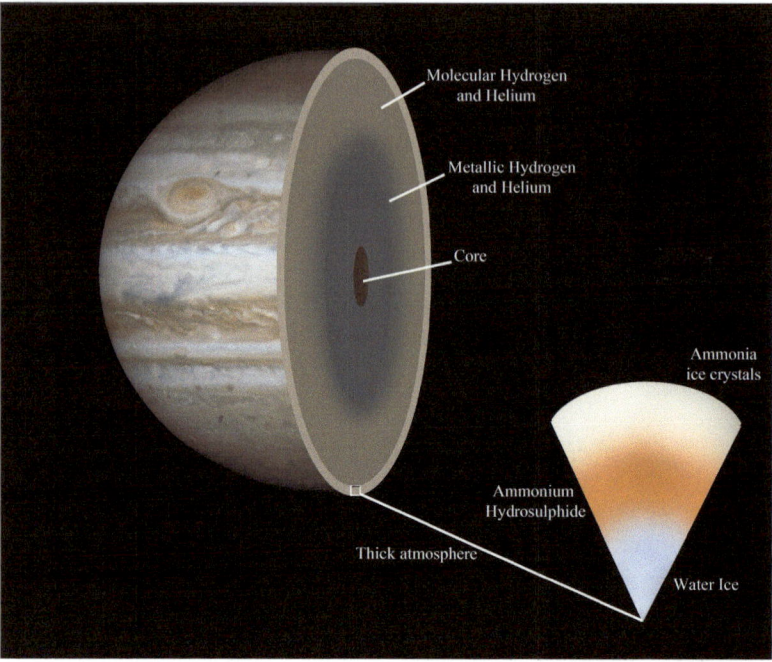

Molecular Hydrogen
and Helium

Metallic Hydrogen
and Helium

Core

Ammonia
ice crystals

Ammonium
Hydrosulphide

Thick atmosphere

Water Ice

Fig. 7.1 Schematic drawing showing the interior of Jupiter

over from when the planet was first formed billions of years ago. The second is the process known as the Kelvin-Helmholtz mechanism. This is where the upper part of a planet (or indeed a brown dwarf star) cools, and so the planet shrinks as a result. The shrinking compresses the core, which in turn generates heat causing the planet to expand again, and thus the planet goes through an endless cycle of heating and cooling. It is thought that this internal heat plays a substantial part in driving the extreme weather conditions on the planet.

Rotation and Longitude

Jupiter is the fasting rotating planet in the Solar System; however, as we noted earlier, its equator rotates quicker than the rest of it. The differential rotation of Jupiter, and the lack of any fixed surface features, presents us with an interesting problem – how shall we define longitude? On Earth this is easily done, as we have a fixed (imaginary) line of 0° longitude that runs through Greenwich (UK), the Greenwich meridian. On Mars, the fixed surface feature Dawes Bay is defined as the place of 0 longitude. But on Jupiter there are no fixed features; even prominent spots and storms move about (sometimes very rapidly).

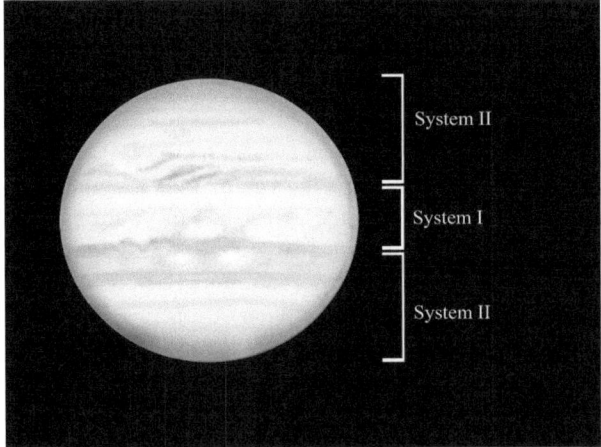

Fig. 7.2 Anything within the equatorial region has a longitude of System I. Everything else is System II

The answer is to define an arbitrary zero that keeps up with the rotation of the planet, rather than choosing a feature. To this end we have two systems of longitude (see Fig. 7.2):

- **System I (ω_1 or L1):** This is the longitude of the equatorial region and has a rotational period of 9 h 50 min 30.003 s. Anything located within this region is given a longitude of System I.
- **System II (ω_2 or L2):** This is the longitude of everything else that is not in the equatorial zone. System II has a rotational period of 9 h 55 min 40.623 s, and so all objects not in the equator are assigned a System II longitude.

There is a further system of longitude, System III. This is the rotational period of Jupiter's magnetosphere, and has a period of 9 h, 55 min. This system is now used by NASA and is becoming increasingly common since it only involves one system of longitude and is not dependent on where the surface feature is located.

When you have made a disk drawing (or a transit timing), you will need to work out what line of longitude is on the central meridian (an imaginary line connecting the north and south poles) at the time of observation. Longitude for all three systems can be worked out in two ways; either very quickly using *WINJUPOS*, or by hand, which takes longer and requires a little calculation.

By far the simplest way is to download the free software *WINJUPOS*. Once you have done this, go to the 'program' menu, then to 'celestial body' and select Jupiter. Next go to the 'tools' tab and select 'emphemerides.' This will bring up a window, and if you input the date and time of your observation, *WINJUPOS* will display the values of CMI, II and III that were on the CM at the time of your observation.

If you don't have access to *WINJUPOS* or prefer to work out longitude by hand, the method is simple. We just use the formula:

$$\lambda_S = \omega_S \pm \Delta t$$

where the term λ_S is longitude of a given system on the CM (for example, System 1 is λ_1) at the time of observation, ω_S is the value of the longitude for the system at midnight on the night of observation. Δt is the time of your observation converted into the appropriate longitude of the system we are working in. If you observed at midnight, $\omega_S = 0$. We use the '−' if the observation was before midnight, and the plus if it was afterwards. Let us look at an example.

Suppose we wish to find the longitude of System II for a drawing of Jupiter made on 2012, July 21, at 0220UT. We shall need to first find the longitude of System II at midnight, ω_S. These values are printed in tables (the BAA *Handbook* is one excellent source). From the tables in the BAA *Handbook* we can read off the value: $\omega_2 = 358.7°$.

If we had observed at midnight, $\lambda_2 = 358.7°$; however we observed at 0220UT, 2 h 20 min after midnight, so we must find out what 2 h 20 min is in System II. From the tables in the BAA *Handbook* we have that:

$$2 \text{ hours} = 72.52°, \text{ and } 20 \text{ minutes} = 12.09°$$

Thus we have that

$$\Delta t = 2 \text{ hours} + 20 \text{ minutes} = 72.52° + 12.09° = \underline{84.61°}$$

Putting this into the equation for longitude we have

$$\lambda_2 = \omega_2 + \Delta t = 358.7 + 84.61 = 443.3$$

This value is greater than 360° (there are only 360° in a circle) and so we subtract 360° giving:

$$\lambda_2 = 443.3 - 360 = 83.3$$

(We only do this if the answer is above 360°.) So at 0220UT on 2012, July 21, the line of longitude 83.3° was on the CM in System II.

The procedure is very similar if we had observed earlier, say at 2348UT. The value of ω_2 is the same, except now we were observing 12 min before midnight, and so we would have to convert 12 min into System II longitude. From the tables, 12 min is 7.25° (so $\Delta t = 7.25$) and so now we would have

$$\lambda_2 = 358.7 - 7.25 = 351.5$$

So this is the value of System II at 2348UT. If the value we had obtained had been negative we would need to add 360° to it to make it positive The longitude value must be between 0° to 360° (inclusive).

Features of Jupiter

If you look at Jupiter at ×150 or more through any decent telescope, it is clear that the disk is not uniform. Jupiter appears to be differentiated into regions of bright zones and dark belts, many of which exhibit fascinating intricate details. The belts are the lower warmer cloud decks, while the zones are the bright ammonia ice crystals higher in the atmosphere. Each of the belts and zones have a name and abbreviation used by amateurs and professionals (see Fig. 7.3). The nomenclature used is the one employed by the British Astronomical Association.

In the discussion below is a brief overview of the various belts and zones and some of the well known phenomena associated with them. The letters in the brackets are the abbreviations currently used by the BAA.

South Polar Region (SPR)

This is the dusky polar hood found in the extreme far south of the planet. Usually a grayish feature, but sometimes traces of brown can be seen here. Often it does not appear to be uniform and may take on a somewhat mottled appearance.

South South Temperate Belt (SSTB)

This is a thin belt immediately below the SPR. The belt can vary in intensity so it is not always prominent. The belt may not always be complete; sometimes it is broken at various longitudes. Numerous white ovals may be found here, although they require a 6 in. (150 mm) telescope or larger to see them.

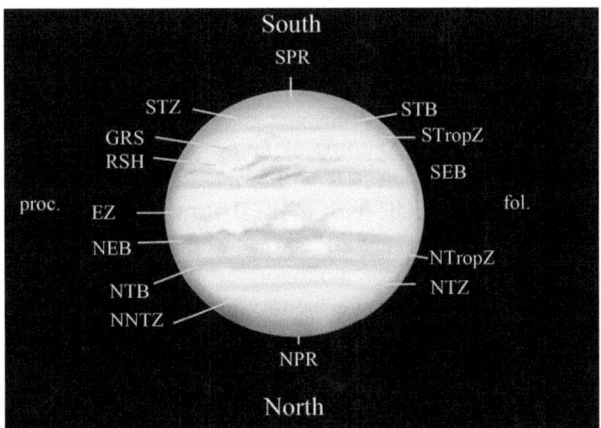

Fig. 7.3 The names of the various cloud belts and zones

South Temperate Zone (STZ)

This is a bright zone, normally white or yellowish in color. A phenomena of great interest in this region are the ovals of the STZ. An oval is (as its name suggests) a bright oval-shaped region in the STZ. In 1939, it seemed that a number of these ovals formed in the region, but only three became definite defined regions. They were designated FA, BC and DE, and it was found that they drifted about in longitude. In 1998, the two ovals BC and DE came together and merged into one single oval called BE. Finally, in 2000, the new oval BE merged with existing oval FA to create a new oval, BA, which is still visible today.

In March 2000, something remarkable happened. Oval BA took on a distinctly orange hue. Christopher Go (a Filipino imager) discovered the color change and other astronomers soon confirmed it. Oval BA is much smaller than the Great Red Spot, and in May 2008, Jupiter's STZ produced yet another red spot (smaller than the GRS and BA), although an encounter with the GRS seems to have demolished it. During the apparition of 2011/12, oval BA lost most of its color, becoming rather obscure. In the 2012/13 apparition, it seems the red color has returned once more.

Obviously we shall have to wait and see what further surprises the STZ has in store for us in the future, but one thing is certain: It is well worth keeping an eye on this region, for who knows what it will produce next!

South Temperate Belt (STB)

This is one of the main belts on the disk. The STB can be quite can conspicuous, and when it is prominent even a 3 in. (76.2 mm) telescope will show it. The STB exhibits a number of interesting phenomena. For one thing, the belt varies in intensity but not necessarily in a uniform manner. Sections of it may become dark and prominent, while other sections become faint and obscure. Occasionally it appears that there are two sections of the belt (so it appears double), or it may become practically invisible at various longitudes.

The STB also seems to be the home of numerous white spots, some of which (when large enough) can be detected with an 8 in. (203 mm) telescope. Other objects that have turned up include dark spots. By timing how long it takes for these objects in the STB to shift in longitude we can determine the underlying wind speeds of the belt. The situation is analogous to dropping a twig into a running stream; if you know how fast the twig is moving then you can work out how fast the current in the stream is moving. The same is true for the features in Jupiter's atmosphere (in this case, the current is the south temperate current). As we shall see later, this work can still be undertaken by amateurs.

South Tropical Zone (STropZ)

This is another bright zone, normally of similar brightness to the EZ. In images and photographs the zone appears white; indeed this is its usual appearance in the eyepiece, although a yellowish hue can be observed on occasions. The zone is notable for its long-term resident, the Great Red Spot and the south tropical disturbance (see below). This is a very active zone on the planet, and well worth close examination during observation.

South Tropical Disturbance

During the 1900 and 1901 apparitions, Bertrand Peek, a former director of the Jupiter section of the BAA reported in his excellent book *The Planet Jupiter*, *an Observers Handbook* that a series of "dark humps" were observed projecting from the southern edge of the SEB into the STropZ. Eventually one of the humps became quite prominent and became connected by a gray filament across the STropZ and a dark streak on the STB. Brilliant white spots developed near the edges of the disturbance, which came to be classified as a south tropical disturbance. The drawing in Fig. 7.4 by David Gray shows the effect. In general with a south tropical disturbance, one finds that there is a darker shaded region in the STropZ. The region is normally quite thin and (usually) has a bright spot at the top and the bottom. It seems that the disturbance normally occurs close to the proceeding edge of the Great Red Spot. Moreover, the phenomena seems to be connected with an SEB revival (see below) since the disturbance seems to occur as a SEB fading is progress.

The Great Red Spot (GRS)

This is probably one of the longest-lived storms on the planet. As its name suggests, the GRS takes the form of a large orange oval normally found in the STropZ (although it can drift a little in latitude it is always to be found in the STropZ). The GRS sits just above the south equatorial belt, but just below it (north of the spot) is an indentation in the SEB called the Red Spot Hollow (RSH). The orange coloring is thought to be the result of complex organic compounds. Over the years, there have been many theories to explain the spot, including a large island floating in the upper atmosphere of the planet. Today, thanks to spacecraft visits we now know that it is a vast hurricane, much larger than Earth.

The GRS was probably first recorded by Robert Hooke (though Cassini made a drawing years earlier that may show it) on 1666, June 26, and it was recorded a number of times after that until there were large numbers of amateur astronomers

Fig. 7.4 Start of a south tropical disturbance as observed by David Gray on 2010, November 15, at 1835UT using a 415 mm Dall-Kirkham at ×270 and ×365 in good seeing conditions. The disturbance takes the form of two bright ovals surrounded by some darker gray material near the CM (Courtesy David Gray)

recording it. From the drawings we have of the spot in the eighteenth and nineteenth centuries, it seems that it was a great deal wider than it is now. Indeed, we now know that the spot is shrinking, and if it continues to shrink at its present rate, it will be circular by 2040. Why this should be the case is not at all clear. The longevity of the spot is probably due to the fact that there is no land mass on Jupiter; on Earth hurricanes start in the ocean and then hit land, causing them to lose their energy. On Jupiter there is no land to hit, so once started, a storm can go on for a long time.

Even before the rise of spacecraft, the GRS spot was known to be a dynamic object. Sometimes the spot is a strong red color, while at other times it can fade to become an obscure feature only seen with larger instruments. Sometimes the spot appears to have a dark border, and telescopes of 8 in. (203 mm) or larger will often show some internal structure. The GRS is not fixed. It tends to drift around in longitude, and measuring this drift allows us to place an estimate on the speed of the jet streams at that part of the planet.

Fig. 7.5 Jupiter with noticeably faded SEB. The drawing was made by the author on 2010, July 10, at 0214UT with his 203 mm Newtonian reflector, ×250 in good seeing

South Equatorial Belt (SEB)

When present (for the belt fades and reappears – see SEB revivals), this belt is one of the most prominent features on the disk. It can be seen in a 3 in. (76.2 mm) refractor, and an 8 in. (203 mm) telescope shows considerable detail within it. The SEB usually consists of three parts; there are two dark belt components – the SEB(n) and SEB(s) – and in between them is a brighter zone called the SEB zone (SEBz). As well as fading and revivals, the color and intensity of the belt changes over time. At the time of writing, the SEBz is a rather dark orange color, but a few years ago it was much lighter. The region near the GRS is of much interest, as there is a lot of turbulence there, and all sorts of intricate structures seem to be present.

An SEB revival may be one of the most stunning events produced by a planetary atmosphere. Although each SEB revival has its own intricacies, the general pattern of a SEB revival can be broken down into the following stages:

1. First, the SEB will start to fade. The strong color will lose its intensity, and the belt will eventually become a faint ghost of itself. This process can take many months to occur, and observers of past events have shown the belt to disappear. However, during the end of the 2009–2010 apparition (and much for the 2010–2011 apparition), it was possible to trace a very weak belt (see Fig. 7.5) where the SEB was.

2. As the SEB starts to fade, the Great Red Spot will become much more prominent. The color (which would have been a light salmon or pink) will become stronger and may become a strong orange color. The spot will be noticeably darker than before. The view of Jupiter when it is like this is really rather striking – a vast white STropZ with a strong GRS being the only main feature in the zone. This will be the picture for a while, a strong GRS and weak (or no) SEB. The exact period of time that this lasts for is uncertain. The last time this occurred, in 2009–2010, the SEB was absent for only a year or so. On other occasions Jupiter has persisted in this state for several years.

3. The SEB revival begins with a single brilliant white spot in the SEB. An outbreak will then occur. This will take the form of a series of white and dark points erupting all the way along the SEB. More storms will appear at the center of the outbreak, and eventually this material will catch up with the GRS. As a result of this interaction, the GRS begins to fade, losing both its strong color and intensity.

4. The final stage of the process sees all of this material distributed by the jet streams along the zone so that the whole belt returns to the appearance it had before it faded. The revival is then over, and for some time the SEB will stay in this form until the whole process is repeated once more.

A less dramatic version of an SEB revival also occurs from time to time. A mid-SEB outbreak starts when a section of the SEB appears to fade (rather than the whole thing). Sometime after the fading, a brilliant white spot will appear. The spot will stretch out in the following direction of the SEB and eventually there will be an outbreak of similar white spots at the location where the first one appeared. Eventually the SEB will become a very chaotic-looking belt, with vast sections of turbulence dominated by light and dark structures following the GRS. Very often it is amateurs who record the start of these outbreaks, so it is well worth being familiar with the current status of the planet so that you can be on the lookout for any sudden outbreaks.

Equatorial Zone (EZ)

This is the bright zone situated at the equator between the SEB and the NEB. Usually it is cream in color. However the southern (or northern part) may take on a yellowish color. Very often a thin grayish equatorial band can be made out within the region. The EZ is often the home to festoons – dark bluish structures that look like loops. They tend to start out just on the southern edge of the NEB and extend into the EZ. They may join the EB or they may not extend quite that far south. Sometimes the EZ has many such festoons; at other times there may be very few. Other fine structures may be present, too; often lighter patches (sometimes yellowish in color) may be present along with dark streaks near the SEB. It has to be said that a large telescope and good seeing is required to follow the festoons in any close detail. When conditions are good, you will find that there is much delicate structure both at the start of a festoon and within the zone itself. Indeed on such nights it can be hard to record the details accurately!

An interesting event occurred during the 1962–1963 apparition. During this time the EZ appeared to darken considerably, so much so that to many observers it appeared that the NEB had merged with the SEB. The drawings of the planet at that time show what appears to be a single enormously thick equatorial belt extending around the whole of the planet. The effect decreased after 1963, and the EZ returned to its usual appearance. The phenomena has not occurred since, and why it happened in the first place is still a mystery. If you have access to a large telescope, a study of the morphology of the festoons and other objects in the EZ will be worthwhile.

North Equatorial Belt (NEB)

This is usually a prominent dark band, just north of the EZ. It can vary a great deal in width; sometimes it is a thin belt (as it is at the time of writing in summer 2012) and at other times it can become quite broad. In the last few years the color seemed to be a sort of orange-brown, but in 2012 it was a little weaker. The belt can sometimes appear to be a double structure with a north and south component. However, unlike its southern counterpart, the zone separating the two sections may be very poorly defined in some sections.

The NEB is the home to a great deal of activity. Its southern edge is often the home to various dark projections, many of which will extend into the EZ in a looping structure which we call festoons. A number of darker sections may also be present within the belt, giving the impression that the belt is comprised of many dark and lighter oval shaped structures and longer elongated sections when the seeing is good.

The northern edge of the NEB (NEBn) is also the site of well established activity. Very often a number of white ovals and dark spots can be seen along the northern edge of the belt. The white ovals seem to be proper structures in that in good seeing (and in images) they stand out against the NTropZ, so they are not simply curved sections of the NEBn. The dark spots can become quite prominent. During the 2011–2012 apparition, there were three large dark spots that appeared to be detached from the NEB (which was fairly thin). The color of these dark spots was quite striking, a dark brown with some burgundy hues seemed to be present (see Fig. 7.6). The spots were not circular and were more like ovals. (Indeed, dark spots along the northern edge of the NEB seem to become elongated over time.)

There seems to be no reason at all why the NEB should not undergo systematic fadings and revivals like the SEB, and yet such events seem to be very rare. Indeed there seems to be only one documented occasion; between 1911 and 1912, the NEB was very weak and had, for all intents and purposes, vanished, according to the drawings made at the time. Then towards the end of the apparition, the belt began a revival when the planet was observed. The following year, it had returned

Fig. 7.6 Drawing by the author of Jupiter showing the two dark areas in the NEB. The drawing was made on 2011, November 1, at 2318UT with a 203 mm Newtonian reflector at ×250

to its usual appearance. During the1904–1905 and 1905–1906 apparitions, the belt appeared to be rather thin (as it does at the moment). However there has not been any dramatic fading or revival of the belt since the 1912 episode. Are we in for a spectacular fading and revival event soon? Quite frankly, no one knows.

North Tropical Zone (NTropZ)

This is usually a bright zone immediately to the north of the NEB. Normally the zone is quite bright, but it's intensity can vary somewhat. The zone usually has a creamy white tone, but this can change. Indeed at the time of writing, the NTropZ is displaying considerable activity and has a strong yellowish-tan hue. Activity wise, the zone seems to be split into two parts; the south part of the zone seems to be the home of the various streaks, spots and ovals associated with the NEBn, while such markings in the northern part of the zone seem to be absent. The occasional bright oval may be observed, but activity in the north is normally quite low (Fig. 7.7).

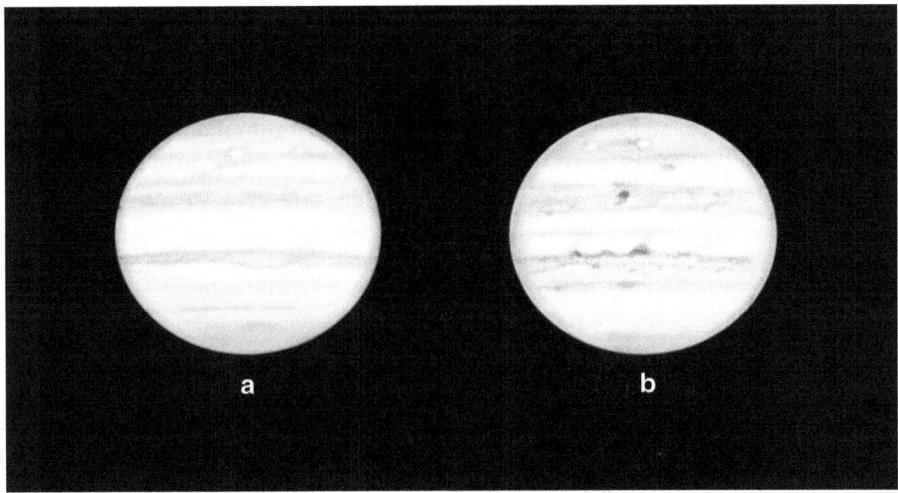

Fig. 7.7 Two drawings of Jupiter made by the author with his 203 mm Newtonian, showing considerable change in the NTropZ and NTB. (**a**) Drawing made on 2011, November 2, at 0131UT, ×250. Here the NTB is virtually absent. (**b**) Drawing made on 2012, August 9, at 0305UT showing considerable activity in NTB. The NTropZ is now rather dark

North Temperate Belt (NTB)

This is usually another conspicuous belt, and when prominent can be seen in small instruments (3–4 in. in aperture). The belt has displayed considerable activity in the past. Sections of it have faded (indeed the whole thing has faded completely a number of times in the past), and there may be parts where the belt is double.

Although the belt may look thin and uniform at a cursory glance, closer scrutiny will often reveal the belt edges to be slightly irregular. The width of the belt also varies, as does the intensity and color. Other phenomena occasionally associated with the belt are the so-called rapidly moving spots. These are a series of dark spots that appear on or close to the southern edge of the belt, and they appear to move rapidly in longitude.

North Temperate Zone (NTZ)

This is another bright zone (although it can take on a dusky appearance), and usually cream in color. The breadth of this zone is subject to change. Sometimes it can be narrow, while at other times it can become quite broad (dependent on where the NTB lies in latitude). Occasional white spots might be glimpsed in the region.

Northern Polar Regions (NPR)

The northern polar region is quite broad. Sometimes the region has a vaguely mottled appearance, and at other times it may be further differentiated into a number of belts and zones. Convention has it that we continue to increase the number of 'north's' in the names of the belts and zones. So if present, the next belt north of the NTB is the North North Temperate Belt (NNTB or N^2TB), which might be followed by a North North Temperate Zone (NNTZ). After the NNTZ would be a North North North Temperate Zone (NNNTB or N^3TB) and so on.

In general the region is fairly quiet, with only the occasional spots and streaks to be found here. On occasions well defined markings have been seen, but in general this is quite rare. The color and intensity of the region is known to vary a little.

The Galilean Satellites

Jupiter has an impressive retinue of attendants (66 satellites is the current count), but of these only four of them are available to amateur telescopes. These are the Galilean satellites, named after Galileo, the first astronomer to observe them. In the older literature you will find them simply designated names Satellite I, II, II and IV; however today, we know them as Io, Europa, Ganymede and Callisto.

These four faithful attendants of Jupiter make a fascinating spectacle as they move about the planet, night after night locked in their eternal gravitational dance. Often one or more of the satellites (and their shadows) will pass in front of Jupiter in an event known as a transit. Other times they will pass behind the planet or into its vast shadow and become invisible for a time. You can identify them quite simply. For example *WINJUPOS* will name the satellites and show you how they will move over the course of a night.

As well as transits, the moons undergo a number of mutual events. Mutual phenomena include the satellites passing into each other's shadows along with mutual eclipses. *WINJUPOS* will provide information about such phenomena, as will most astronomy magazines and books.

The Galilean satellites were once thought to be flotsam, ancient dead worlds locked into eternal cold and of no interest. This belief has been completely overturned! These moons have turned out to be some of the most interesting and geologically active places in the Solar System! Whole books could be devoted to the study of them. Here, however, we shall restrict ourselves to a passing survey of them. We start, with Io.

Io

This is the innermost of the Galilean moons. Its diameter is around 3,642 km, and as such this makes it larger than our own Moon. Io is unique for being one of the most

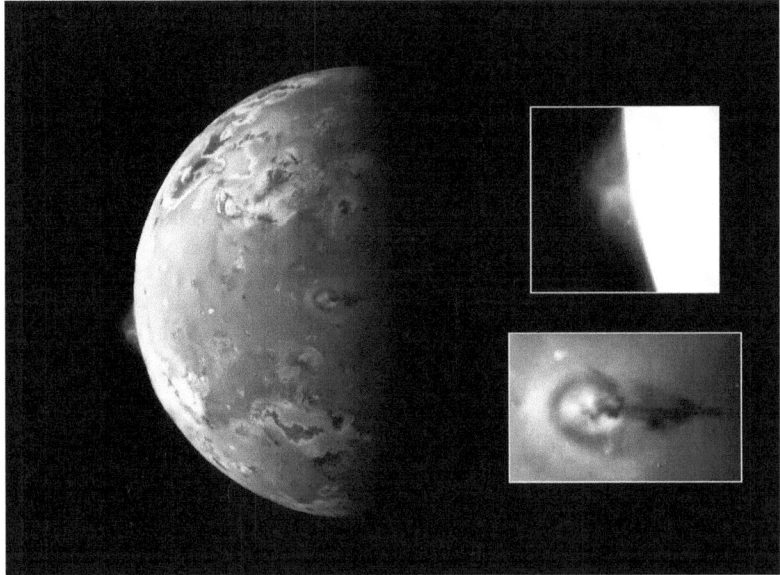

Fig. 7.8 A wonderful image from the Galileo spacecraft showing the volcanic surface of Io. The plume that can be seen is from a volcanic caldera called Pillan Patera. The plume extends to some 140 km above the surface (Courtesy of NASA JPL)

volcanic bodies in the Solar System! The surface is covered with swirling layers of sulfur that have been spewed out from the many volcanoes on the surface. Although the surface temperature is relatively cool ($-143\ °C$) the temperature of the volcanoes is rather high ($2,000\ °C$ as measured by the Galileo spacecraft). For all this volcanic activity to exist, there must be a great deal of heat being generated inside the moon, and the cause of that heat is clear. Io's orbit is not perfectly circular; one side of Io faces Jupiter, and so the strong gravitational pull of the planet tugs the surface, while the other side is being pulled by the other Galilean moons. This means the interior of Io is being flexed and pulled, and this is the source of the heat (Fig. 7.8).

Internally the moon probably has a rich iron core that may extend quite some way from the center (possibly halfway up to the surface). An interesting phenomena associated with Jupiter and Io is the so-called Io flux tube. Io has a tenuous atmosphere of sulfur dioxide. This, along with the material erupting from the volcanoes gets swept up by Jupiter's vast magnetosphere. There is a large cloud composed largely of neutral atoms of sulfur, oxygen and other elements surrounding Io. The atoms of this cloud can become ionized, and when this happens they are influenced by Jupiter's rotating magnetosphere. This effect has created the Io plasma torus, a doughnut-shaped region of charged atoms. The cloud and Io's atmosphere are connected to the magnetic field lines of Jupiter, and when the charged particles flow to the north and south pole they create a substantial electric current. This creates tremendous aurorae at the north and south poles of the planet.

Fig. 7.9 Image of Europa from the Galileo spacecraft taken from approximately 725,000 km above the moon. The surface of Europa is covered in an intricate web of lines caused by cracks and fissures in the ice (Courtesy of NASA)

Io just about always appears as an orange color to me, though seeing it that way may have come from seeing the splendid images of the Voyager and Galileo probes. Interestingly enough, the color of Io was the subject of debate many decades before the first spacecraft visited the satellite. Patrick Moore in particular suggested the moon had a distinctly orange color associated with it. When Io passes across the face of Jupiter, it can become quite hard to see and appears a grayish-yellow color. During such transits it is interesting to see just how quickly Io moves as it shuttles back and forth in its orbit.

Europa

This is the next satellite out from Io and is the smallest of the four, at around 3,120 km in diameter. Europa couldn't be more different from Io; there are no large volcanoes here. Indeed the surface is very smooth. Europa is completely covered in ice, which in places has a similar appearance to frozen sea ice here on Earth. Indeed, with relatively few impact craters on the surface, it is clear the ice must be relatively young (in a geological sense), probably only a couple of millions of years old, so some action must be responsible for this resurfacing (Fig. 7.9).

Europa has been the subject of much speculation in recent decades as a possible candidate home for extraterrestrial life. The same gravitational heating that powers the volcanoes on Io also affects Europa, but much less. It is thought that the surface ice may extend down to a depth of between 10 and 25 km, but after that, it is has been suggested that the ice has been melted by tidal heating to form a vast ocean maybe as much as 100 km deep. Results from the Galileo probe seem to imply that this ocean is quite salty, and this has consequences for the development of life.

Do strange alien beings swim in the dark oceans of Europa? Such a biosphere would be protected from the cold harsh conditions above by an eternal, impenetrable ceiling of ice.

Like Io, Europa will also transit Jupiter. When Europa passes in front of a bright zone it can be very difficult to observe, however.

Ganymede

This is the largest Moon in the Solar System, and with a diameter just over 5262.4 km it is larger than the planet Mercury. Ganymede is very different from Io and Europa. Its surface is characterized by bright and dark regions; the darker regions are quite heavily cratered, so this suggests it is older. Formations of fractured circles also populate the darker regions. The brighter ones seem to have strange intricate groves that can span many thousands of km along the surface of Ganymede. It seems likely that the satellite bore witness to a good deal of bombardment in the early days of the Solar System (Fig. 7.10).

Like Io and Europa, Ganymede will also be the subject of tidal heating, but this will be much less powerful than it is for Europa. Nevertheless, it is believed that deep under the surface (maybe as much as 200 km) there is a salty ocean. Interestingly, the Galileo spacecraft found that Ganymede had a permanent intrinsic magnetic field that is three times stronger than that of Mercury's. This magnetic field marks out a well defined (but small in astronomical terms) magnetosphere, and this is the only satellite in the Solar System with this property.

All of the Galilean satellites appear as small disks in telescopes with apertures if 6 in. (150 mm) or more but Ganymede is of particular interest because it is the largest, and has an average diameter of 1.6". Imagers like Damian Peach have been able to record some details on the disk but can anything be seen visually? The answer seems to be yes. Some details can be detected with large amateur telescopes.

Ganymede transits Jupiter, but less frequently than Europa and Io. Ganymede is a grayish brown color as it passes in front of the bright zones. It can be harder to see when it grazes the polar sections. One thing that cannot be missed is its shadow, which is startling in size and is as black as ink.

Fig. 7.10 A striking picture of Ganymede from *Voyager 2* (Courtesy of NASA)

Callisto

The final Galilean satellite is remarkable in the fact that is completely unremarkable. Callisto is the second largest of the four, just over 4,820 km in diameter. Io, Europa and Ganymede all experience orbital resonance. Resonance is the effect where the orbital period two bodies are related by a small whole number. For example, Europa has a 2:1 ratio resonance with Io; for every lap Io makes, Europa completes two and so on. This resonance is a source of tidal heating, but interestingly Callisto does not have an orbital resonance with the other satellites. Nevertheless, models of the moon's interior suggest that there may well be a small ocean deep down under the crust. The ocean is likely to be a relic of the slow convection that started up inside the satellite not long after it had formed (Fig. 7.11).

The surface of Callisto is very dark and heavily cratered, and it is the faintest of the four moons. The surface seems to be very ancient and in many respects is similar to Earth's Moon (although much more heavily cratered). Callisto undergoes transits, but because of its long orbital period (nearly 17 days) these are much rarer.

Fig. 7.11 Galileo image of Callisto showing the dark, heavily cratered surface (Courtesy of NASA)

Historical Accounts and Observations

Jupiter is one of the brightest objects in the night sky, and there can be no way the ancients could have missed it. Being such a nocturnal beacon, it is not surprising that Jupiter became one of the first targets for the early telescopic explorers, the results of which provided a radical shift in humankind's view of how the universe worked.

Galileo was the first telescopic observer of Jupiter to record his observations. On the night of the 1610, January 7, he turned his small telescope towards Jupiter. To the naked eye, the planet appeared as a bright yellowish star, but in his telescope (which probably only had a power of ×30) the view was resolved into a small bright disk. Near to Jupiter were 'three fixed stars.' Who knows what Galileo thought as he watched for the next few nights as the three stars changed their positions about the planet. On 1610, January 10, one of them seemed to have vanished, and on January, 13 he discovered a fourth bright object. No other star in the sky behaved in the way that these four did, and there was only one conclusion: these were moons of Jupiter, orbiting the planet. The one that had disappeared had simply passed behind the planet.

This had a profound effect on Galileo and his discovery of the moons of Jupiter (which we now call the Galilean moons) marked a new dawn in astronomy.

The prevailing view of the workings of the cosmos as Galileo made his observations was the one proposed by Aristotle (384–322 B.C.). In Aristotle's model of the universe, the Sun, Moon and planets all orbited Earth in circles, which was located at the exact center of the universe. Earth was the most important object in the universe – this was the reason why everything orbited it. The Catholic Church in particular was rather fond of this model, and it became enshrined in scientific and religious thinking (which were not so clearly distinct from each other as they are today).

Galileo was troubled by the question, why were the moons of Jupiter not orbiting Earth? He continued to observe the moons and obtained quite reasonable estimations of their orbital periods. Galileo would eventually champion the correct model of the Solar System with the Sun (rather than Earth) at its center. He was treated rather badly by the Church and put under house arrest for the rest of his life for revealing the truth about the world we live in.

As we have seen in earlier chapters, the early telescopes were small and clumsy affairs, but the astronomers that used them were equipped with fine minds and perseverance that more than made up for the shortcomings of their telescopes. The next two observers to make a significant contribution to the study of Jupiter were the Italian Giovanni Cassini and the English scientist Robert Hooke. Both were observing Jupiter from the late 1640s onwards. Although the telescopes they were using were small, the atmosphere of Jupiter produced bright ovals and spots, features that tantalized these early explorers. Cassini was probably the first person to observe the Great Red Spot. From 1648 to 1669 Cassini was professor of astronomy at Penzano University, and in 1671 he became director of the Paris observatory. At the time, both he and Hooke were observing a bright oval in Jupiter's STropZ, which was probably the Great Red Spot. Cassini used the spot to obtain an estimate of Jupiter's rotational period. However he found that if he used objects in the EZ for the same measurement, there was a discrepancy of some 6 minutes. Cassini had discovered that Jupiter had differential rotation; its equator rotates quicker than the rest of it.

Other astronomers began to observe the features we are familiar with today. Shröter and Madler observed bright and dark spots, as did Dawes. Unlike the uncertain views of Venus, or the challenging tiny disk of Mars, Jupiter's vast surface offered striking and ever changing details and patterns. In 1878, it seems that color returned to the Great Red Spot. Previously, in 1869 and 1870, observers had noticed a dark ellipse in the STropZ. In July 1878 another ellipse appeared in the same region as the one of 1869–70, only this time it contained a very strong red color. This astounded observers at the time; no other planetary surface had produced such a startling object. Drawings of the time paint a remarkable picture. It seems that the GRS was much wider and redder than it is today (indeed we know that the spot is slowly shrinking). Jupiter simply provided far too much work for one person to study, and if anyone was ever going to make sense of the planet, clearly an organized systematic effort would be needed.

In 1890, a movement was underway. Many amateur astronomers were dissatisfied with the Royal Astronomical Society. It was costly and, moreover, it did not allow females to became members. Many astronomers came together and formed the British Astronomical Association. A number of observing sections were formed to study specific astronomical objects. The Jupiter section in particular was one of the early sections to form, and members soon started work collecting observations. In 1958, then director Bertrand Peek published his splendid book, *The Planet Jupiter: The Observer's Handbook*. The book was almost entirely based on the observations of the Jupiter section and provided practical advice about observing the planet along with presenting the established facts at the time. More recently, current Jupiter section director Dr. John Rogers has also published a book, *The Giant Planet Jupiter*, which, like Peek's book, presents updated facts and methodology.

There can be no doubt that the planet Jupiter offers the amateur astronomer with a vast number of opportunities to contribute to astronomy.

Observing Jupiter

In spite of its vast size and brightness, Jupiter can be a tricky object to observe simply because there are so many details visible at any one time, and moreover these features are being carried very quickly by the planet's rotation. A small telescope of 3 in. (76.2 mm) will present a flattened disk and show the two main equatorial belts along with the four main satellites. Good views can be had in a 4 in. (101.6 mm) telescope, but to undertake serious work a telescope of 6 in. (150 mm) or more will be required. You should be able to get some truly splendid views of the planet with an 8 in. (203 mm) reflector, a size quite satisfactory for planetary work.

Magnification wise, ×150 is probably the minimum power required to undertake most of the projects discussed below. Powers of ×167 to ×250 are good for average seeing, and in good seeing a power of ×300 or will reveal some stunning intricate details. Below is a list of some projects that can be undertaken by the amateur. Before embarking on any of them (especially the drawing), though, it is best to spend some 15 minutes observing the planet and taking in the details. This period can also be used to record transits (see below) and so establish in your mind what is present on the disk. If you start work straightaway, you're bound to notice features you missed because your eyes had not yet captured all of the details on the planet.

Advanced Projects

Disk Drawings

A disk drawing of Jupiter is perhaps one of the hardest things in visual planetary observing to execute. There are two main problems:

1. The vast number of features visible on the disk, especially when the seeing is good can be overwhelming.
2. The swift rotation of Jupiter moves the features very quickly. Indeed there is a small but appreciable drift after 6 minutes.

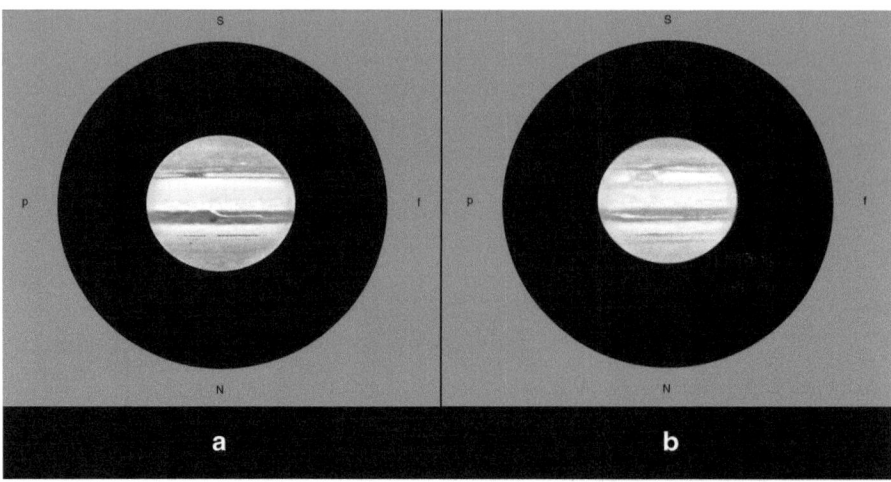

Fig. 7.12 Two color drawings by American observer Carlos Hernandez with his 228.6 mm Maksutov-Cassegrain. (**a**) This drawing was made on 2010, October 2, at 0345UT, ×310. It shows a much faded SEB, and considerable fine detail in the NTB, STB & SSTB. (**b**) This was made on 2010, September 19, at 0330UT, ×310. The GRS is passed the CM (Courtesy Carlos Hernandez)

The first thing we need when making a Jupiter drawing is a correctly sized blank. It is obvious even at the most cursory glance at Jupiter that the planet is not a circle; rather it is decidedly elliptical, and so our planetary blank must accurately record this. The Jupiter section of the BAA uses a blank of 64 mm diameter by 60 mm in height, and this is a good scale to use for Jupiter drawings (see Appendix).

When you come to make a disk drawing, make sure the magnification you use is suitable for the seeing conditions. When you are certain of what is on the disk, you should start your drawing. Remember you only have 12 minutes to make a drawing; any longer than this and Jupiter's rotation will have moved the features out of sight and your drawing will be inaccurate. It is probably a good practice to draw in the main belts that can be seen first, then spend 8 minutes putting in all of the fine details within the belts and zones. You should put in the details on the proceeding edge of the disk first, as these will soon be carried off by Jupiter's rotation. Once this is done the time (in UT) should be recorded, then spend the last 4 minutes putting in the fine details you can see outside of the main features. It is also a good idea to add notes about color and intensity so that you can recreate everything as you observed it when you come to make the final, neat copy.

Color drawings are always nice to make, but you should remember that color is a very subjective thing, and color sensitivity varies from person to person. It is probably best to make color drawings using pastel colors, as they feel more natural, but some people, such as Carlos Hernandez (Fig. 7.12) and David Gray, use computer packages to color their drawings. If you do decide to make color drawings, it is best to make them indoors from your observing notes rather than at the eyepiece.

Transit Timings

A transit timing is the time that a specific feature was situated on Jupiter's central meridian. Once a time is obtained, it can be converted into longitude. Features on Jupiter are not static; they drift about in longitude, with their drift rate determined by the jet streams they are situated in. So, by measuring the drift rate of a feature we can estimate the wind speed of the jet stream affecting it. The aim is to get as many drift rates for as many different features as possible. (This isn't difficult when Jupiter is well placed and seeing is reasonable!)

It is best to record your transits in a table. Below is an excerpt from one of the author's Jupiter observations showing transit times for the night.

2012, August 9

In the first column we have a number. It is best to give your features a number so you can keep track of them. Next we have a "Feature" column. Here we give a brief description of the feature that transited. To save space abbreviations are used such as 'd' for dark and 'w' for white and so on. The next column is for recording the time of the transit, while the two columns after that are for longitude. If the feature is on the equator then it gets the L1 longitude; otherwise it is L2. Finally we have a column for any remarks about the feature. When Jupiter is around in the winter months you will be able to record many hundreds of transits over an apparition. Even on an average night, it is possible to get quite a few transits as can be seen in Fig. 7.13 – an excerpt from one of Sir Patrick Moore's Jupiter books.

At the end of the apparition, you can take the transits of each feature and plot a drift chart. Drift charts tend to be plotted with date along the vertical axis, and longitude along the horizontal axis. Once the points of longitude for each object are plotted, it is easy for the eye to see how the feature drifted with longitude.

Full and Half Rotation Drawings

If you are going to observe Jupiter for a long period of time (say 4 hours or more) it is probably better to make a global strip map. Remember that Jupiter rotates once in just under 10 hours, so if you have observed for 5 hours you have observed more than half of the surface. For northern hemisphere observers, when the planet comes to opposition in the winter months (December, January, February), the whole planet

No.	Feature	UT	L1	L2	Comments
1	Bright oval in EZn	0248	303°	...	Seeing in W#11 filter
2	D. spot in SEBz	0251	...	75.6°	"
3	D. festoon in EZ	0301	310.9°	...	IL
4	W. oval in STB	0305	...	84.1°	IL

1967

<u>March 10</u>. Cleared from 22-23^h, and I used the 12½ in., but seeing was so excruciating that it was really pointless. Clouds then came up, and that settled matters finally.

<u>March 11</u>. Began with 8½ in. refl: v-poor

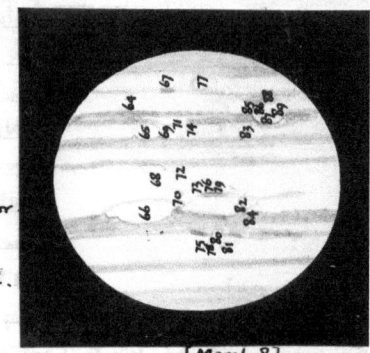

[March 8]

1090.   19.17.   Bright white spot in STZ.
                    ω^2 = 231.5.

Clouds at 19.25. Returned to 10 in, with Oliver, later.

			ω'	ω^2	
1091.	20.20.	Projection from NEBs.	298.2	...	10 in ×360
1092.	21.03.	White spot in NTrZ.	...	295.6	?
1093.	21.04.	C. of long dip in NEBs.	325.0	...	?
1094.	21.43.	Major projection from NEBs.	348.6	...	
1095.	21.45.	P. of white spot in STrZ.	...	321.0	Rain, 21.51
1096.	23.15.	P. of white spot in STZ.	...	015.4	Clouds – patchy
1097.	23.17.	P. of Red Spot.	...	016.6	Dubious
1098.	23.20.	Projection from NEBs.	047.9	...	
1099.	23.31.	C. of white spot 1096 in STZ.	...	025.1	
1100.	23.34.	C. of Red Spot.	...	026.9	
1101.	23.41.	F. of white spot 1099 in STZ.	...	031.1	
1102.	23.53.	F. of Red Spot.	...	038.4	Rain, 0^h
1103.	00.16	[March 12]. F. of dark streak, N. edge NEB.	...	052.3	
1104.	00.19.	P. of white spot in NTrZ.	...	054.1	Clouded.

Seeing horrible – periodic rain! The white spot has now overhauled the Red Spot. Transits 1097-1104 confirmed by Oliver.

Fig. 7.13 One of the pages in Patrick Moore's Jupiter book. Note how many transits can be made on a good night! (Courtesy Patrick Moore)

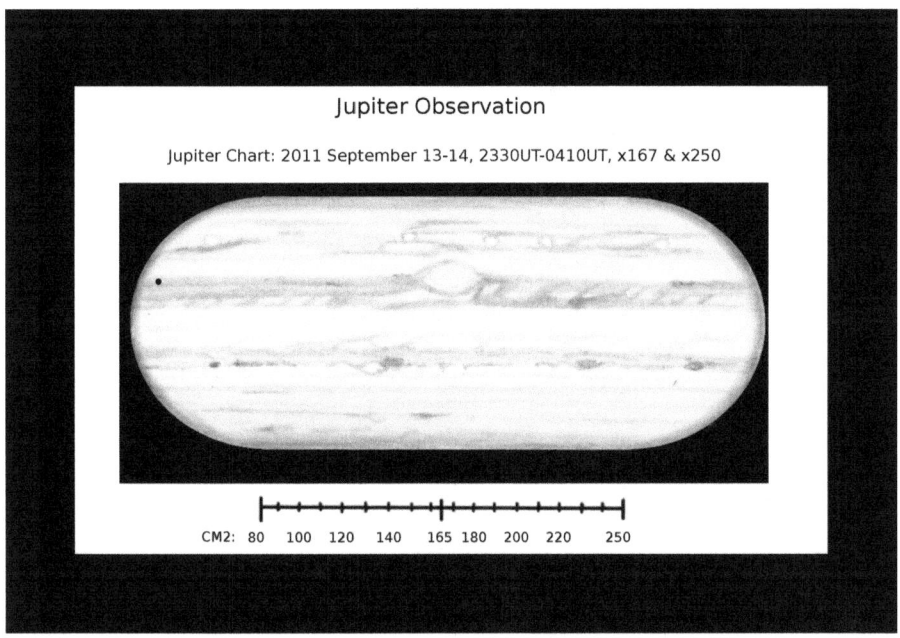

Fig. 7.14 Half rotation map of Jupiter made by the author on 2011, September 13–14, with a 203 mm Newtonian reflector

can be covered in one night, and a full surface map can be made (included are half and full rotation blanks in see Appendix). A half rotation strip map made on 2011, September 13–14, is given in Fig. 7.14.

To make a strip drawing, simply start as you normally would, but fill in the proc. side of the blank first up to the CM. Then as time progresses, draw in the features as Jupiter rotates. Make transit estimates at the same time, and then a CM scale can be added to the drawing afterward (see Fig. 7.14) to show where features are situated. Of course, the map will only be accurate for a few days at most (due to the turbulent and ever-changing Jovian atmosphere), but it is very satisfying, not to mention challenging, to draw the complete surface of another world in one night!

Quite often, a small section of the planet will throw up a very interesting detail, and you might wish to make a drawing just to illustrate this. For this you can use a small sectional drawing. To make a sectional drawing, find the highest power that the seeing will tolerate, and spend about 10 minutes just drawing that small section of the planet. It is best to draw it in a box rather than on a planetary blank. You can fill in additional details such as intensity and color. Figure 7.15 is a sectional drawing made by Carlos Hernadez of the GRS.

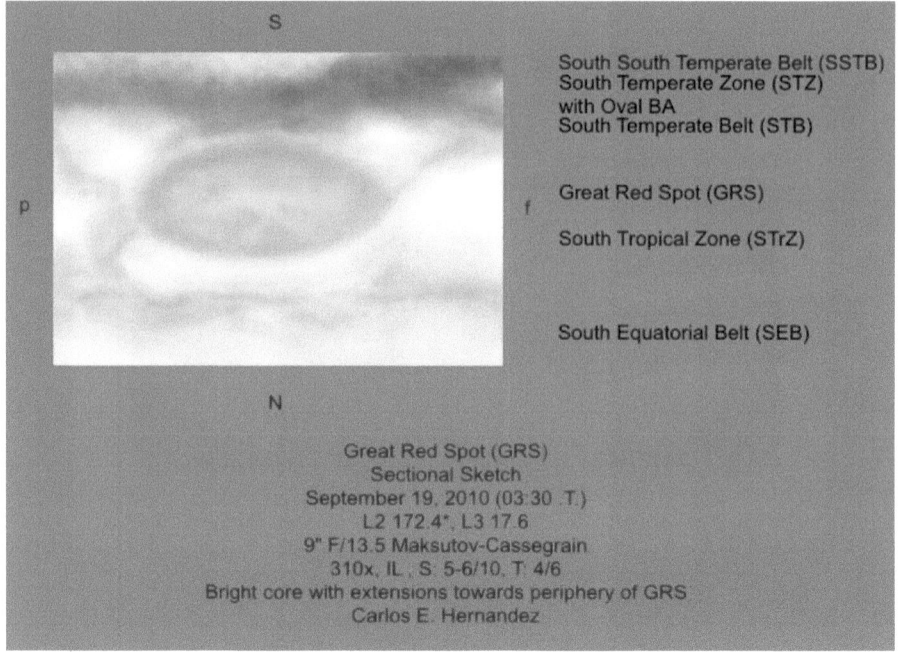

Fig. 7.15 Sectional drawing of the GRS and surrounding environs as observed by Carlos Hernandez on 2010, September 19, at 0330UT with a 228.6 mm Maksutov-Cassegrain ×310. Subtle details in the GRS and surrounding regions are nicely represented (Courtesy Carlos Hernandez)

Sometimes a dramatic feature may change as the night progresses, in which case a number of section drawings are useful to show these changes over time. You should keep the dimensions the same so direct comparisons can be made.

Intensity Estimates

This is a measurement of how bright or dark a feature appears on the Jovian disk. The BAA uses a scale of 0 (brilliant white) to 10 (black sky). Half (0.5) and quarter (0.25) measurements may also be used. To record your intensity estimates, you need a clear bright image in the eyepiece (never make them if the seeing is poor). Next, in your log book, list the features systematically (from south to north, or vice versa) you can see. Then, next to each of the features, assign a value from the scale. Recently, the SEBz was showing as fairly dark at 6.25, while the EZ was rather bright at 1.00, and so on.

Once you have collected an apparition's worth of data, you can see how the features varied over the course of the year. When you have covered many apparitions, you will have a lot of data covering the variation in intensity of many features over the course of many years!

Filter Work

Jupiter responds very well to filters, largely because it is so large and bright! Here are the recommended filters for Jupiter:

- W#38A (dark blue), W#80A (light blue). These are good filters for increasing the contrast of any reddish features (such as the equatorial belts and the Great Red Spot). They also help to enhance the white ovals that appear in the SSTB.
- W#25A (red), W#23A (light red), W#21 (orange). These are largely red filters, and so they increase the contrast of bluish-colored objects. The festoons that come and go in the EZ have a bluish component, and so these filters will help to enhance them (useful for transit timings!).
- W#11 (light yellow). This is helpful for increasing the contrast of the reddish storms situated within the main equatorial belts.
- W#56 (light green). The temperate belts often contain a brownish color, and this filter helps to enhance features that are this color.

There are a number of investigations that you can undertake with the above filters. First, you can make intensity estimates with them, for the features will have different brightnesses in different filters. It is best to be systematic about this; there is little point in making intensity estimates with many different filters from night to night; rather, choose two or three and stick to them for the whole apparition. Try using a W#25A (red) and a W#80A (light blue) over a whole apparition to make intensity estimates alongside the IL estimates.

Another thing you can do is make disk drawings or sectional drawings of features using a specific filter. You might find that, for example, oval BA is better represented in a blue filter (as it sometimes contains an orange-red color), so a strip drawing made with this filter may be better than one made in integrated light. Indeed, there are many storms and structures within the main belts and zones that respond best to different filters, and if they are of particular interest, you should find the filter that brings out the most detail and keep track of any changes to the feature using the same filter. If you haven't got any filters, get some. They will really increase the number of investigations you can make.

Comet and Asteroid Strikes

It used to be thought that planetary asteroid strikes were quite rare; indeed, up until 1994 no one had ever observed such an impact. This all changed in 1994 when Comet Shoemaker-Levy 9 collided with the planet in the most spectacular fashion. The comet was discovered by Gene and Carolyn Shoemaker, and their observing partner David Levy. The trio were photographing the sky using the 0.4 M telescope in Palomar, California, when quite by accident, they photographed the comet. They realized very quickly that the comet was unusual since it was elongated and showed multiple nuclei. The late Brian Marsden at the Central Bureau for Astronomical Telegrams made the calculations and determined the comet would collide with the planet in July 1994, which it duly did. During the course of a week, 26 fragments collided with the planet. The largest hits left black impact marks on the planet that were visible in the viewfinders of amateur telescopes!

After Shoemaker-Levy 9, it was generally agreed that this was a once in a lifetime event (which for comet collisions of this magnitude, it probably is). However, on 2009, July 19, Australian observer Anthony Wesley was imaging the planet when he observed a black spot on the planet. This was later confirmed to be an impact site.

On 2010, June 3, Wesley (and independently, Filipino astronomer Christopher Go) detected yet another collision. This time, however, only a flash of a few seconds was observed on the SEB and no impact mark was to be found. The impactor was probably a small asteroid or comet, but it shows that collisions with Jupiter may not be as rare as previously thought. Much more recently Dan Petersen observed a bright flash on 2012, September 10, other observers caught the impact on video so it seems that this was yet another genuine impact event!

The flashes recorded by Wesley, Go and others would certainly be detectable visually at the eyepiece, so you should keep a constant lookout for further events. If you do suspect such an event, report it to the BAA Jupiter section at once! These events are quite unpredictable, and no one knows when the next such impact will happen.

Out of all of the planet's in the Solar System, you will find that is Jupiter, which will keep you very busy. The planet is an excellent training ground for the visual observer.

Chapter 8

Saturn

Distance from the Sun: 1,513,330,000 km (max.), 1,353,570,000 km (min.)
Diameter: 120,536 km
Sidereal Period: 29.66 years
Synodic Period: 378.09 days
Rotational Period:
   System I (Equator): 10 h 13 min 59 s
   System II (Everything Else): 10 h 38 min 25.4 s
   System III (Magnetosphere): 10 h 32 min 35 s
Mean Surface Temperature: −180 °C
Apparent Diameter: 20.1″ (max.), 14.5″ (min.)
Albedo: 0.47
Brightest Apparent Magnitude: −0.3
Satellites: ~200

A Brief Look at Saturn

Saturn is be one of the loveliest objects the human eye will ever beheld. A small telescope will reveal a scene both alien and mystical; a small brilliant disk surrounded by an unlikely ring system, the planet's many moons are like a string of pearls floating serenely about their parent world.

P.G. Abel, *Visual Lunar and Planetary Astronomy*, The Patrick Moore
Practical Astronomy Series, DOI 10.1007/978-1-4614-7019-9_8,
© Springer Science+Business Media New York 2013

In spite of its ornate beauty, you will hear seasoned amateurs describe the planet as bland; a quieter version of Jupiter is often how the planet is discussed. Although it is true that Saturn's features are more muted, the patient dedicated observer will see subtle changes in the colorful cloud tops, as well as occasional storms and other long-term phenomena both on the disk and the rings. For the visual observer, Saturn is an investment. It may be many years before the planet's atmosphere produces anything dramatic, but when it does, the effects (as we shall see) can be remarkable.

Orbits and Apparitions

Saturn takes just under 30 years (29.66 years) to complete one orbit around the Sun, and so the planet takes a much more leisurely pace and comes to opposition 13 days later each year. Saturn may spend a number of years in any of the ecliptic constellations. This is good (for northern observers) when the planet is in the winter constellations of Taurus and Gemini, but not very helpful when it is in the low summer constellations of Scorpius and Capricornus!

Saturn has an axial tilt of 26.7° The rings are situated around the equator of the planet (called the ring plane), and so our view of them changes as Saturn moves through its orbit. We measure the apparent observed tilt in degrees, and you will find on most visual report forms of Saturn (and in *WINJUPOS*) that this value assigned to the letter B. Thus at $B = +28°$, Saturn is at its southernmost presentation, at $B = 0°$, the planet is edgewise on and at $B = -28°$, the planet is at its northernmost tilt.

In Fig. 8.1.is a rough drawing showing Saturn in its orbit and the view we get of it from Earth. Let's start at Position 1, $B = 0°$. Here the rings are edgewise to us and appear as a thin line; in small telescopes they may even become invisible (especially if the planet is viewed when the Sun and Earth are in the ring plane). As Saturn moves further its orbit, more of its northern hemisphere is shown, and B increases and will pass through aphelion until we reach Position 2, where the planet is at its northernmost presentation ($B = -28°$).

For northern hemisphere viewers, the planet appears low down in the summer constellations at this time. After this the planet's southern hemisphere starts to point towards us again (Position 3) until we come to another edgewise presentation (Position 4). The planet will continue to increase its southerly presentation until it reaches Position 5 (the maximum southern presentation) at $B = +28°$. During this time, the planet will be (for northern hemisphere viewers) high in the winter sky. The rings will be wide open, and the planet is a magnificent sight in any telescope! After this, Saturn continues in its orbit past perihelion, and now the planet's northern hemisphere will start being presented (Position 6) until we move back to the edgewise presentation at Position 1 and the whole cycle repeats once more.

Due to the nature of Saturn's orbit, the time between successive edgewise presentations is not equal. In fact, it takes 15 years 9 months to pass through the aphelion one, (that is, from $B = 0°$ to $B = -28°$ and back to $B = 0°$ again) and just 13 years 9 months to pass through the perihelic one (i.e., $B = 0°$ to $B = +28°$, and back once more to $B = 0°$).

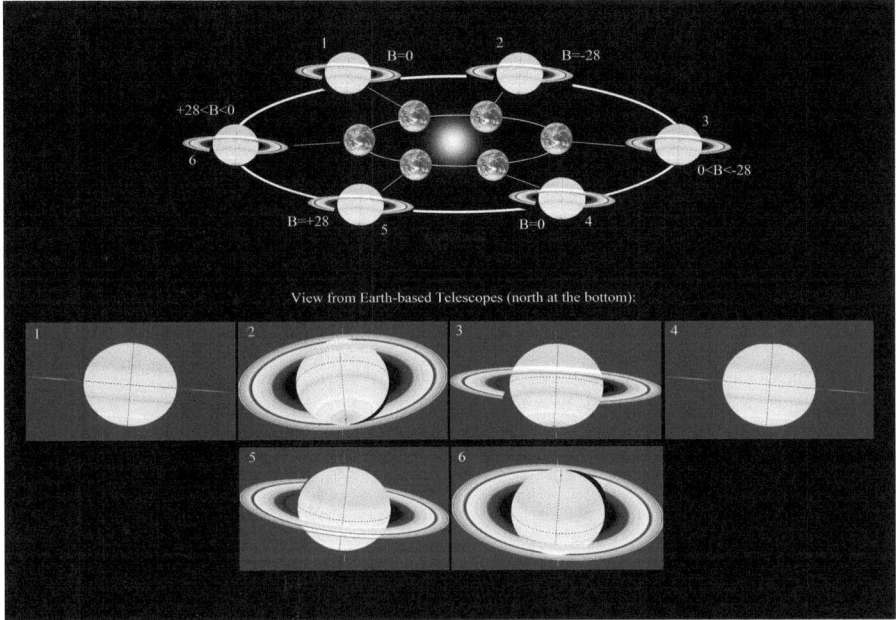

Fig. 8.1 Saturn's orbit around the Sun and the changing aspect of the planet as seen from Earth. The various values of *B* are given

Saturn, the Gas Giant

Like Jupiter, Saturn is a gas giant and in Fig. 8.2 is a diagram that gives the interior of the planet as we currently believe it to be. At the very top, we have a thick and turbulent atmosphere. It is currently believed that the atmosphere consists of three cloud decks. At the top we have the cold clouds of ammonia droplets and crystals. Here the temperature is −250 °C. Beneath this layer we have a second cloud deck composed of ammonium hydrosulfide and is at a temperature of −70 °C. Finally we come to the last cloud deck, which is composed chiefly of water and is at a temperature of 0 °C.

As we push down further beneath the atmosphere we come to a vast region of molecular hydrogen and helium. Even further down, Saturn's great internal pressure squashes the hydrogen and helium into a form of metallic hydrogen and helium. As with Jupiter, this swirling sea of metal provides Saturn with a powerful magnetic field. Beneath this at the center of the planet it is believed there is a hot rocky-silicate core, although of course for obvious reasons, it is very difficult to investigate the interior of the planet.

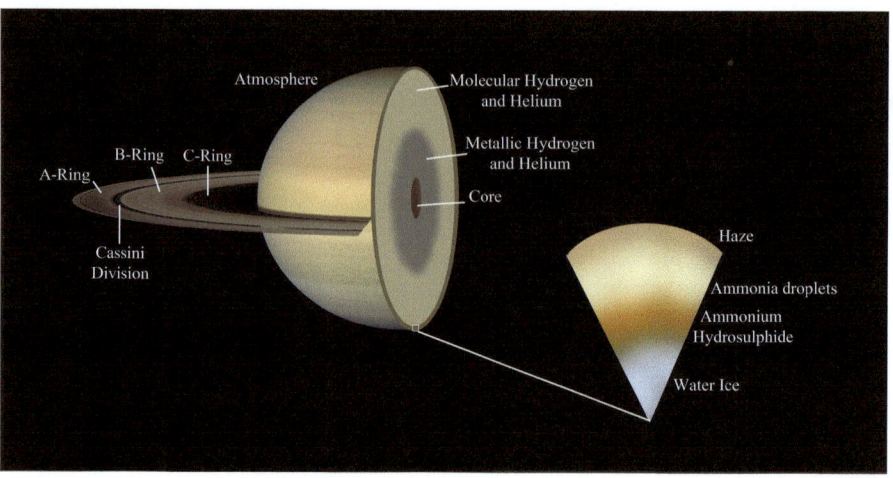

Fig. 8.2 Internal structure of Saturn

The Rings

The vast ring system for which Saturn is so justly famous for are not solid. It was the English mathematician James Clerk Maxwell (famous for his work on electromagnetism) who, in 1859, showed that a solid ring would be pulled apart by Saturn's gravitational field (since the inner part would orbit Saturn quicker than the outer part of the ring), and if it was composed of a fluid, the ring would be reduced to a series of blobs and pieces. In fact we now know that the rings of Saturn are composed of millions of particles ranging in size from small grains of sand to boulders the size of a house. In width the rings are extensive. The E-ring goes out to some 300,000 km, however, in terms of thickness, the rings themselves are very thin.

The main rings themselves are composed of thousands of smaller rings, and the whole system is very complex and dynamic. We have large divisions where there is little material (such as the Cassini division) and shepherd moons that seem to keep some rings in place. The whole structure is one of the most beautiful things to exist in nature.

It has to be said, the origin of Saturn's rings is somewhat of a mystery, the leading theory is that a moon perhaps strayed too close to the planet and was broken up by Saturn's gravitational field. This material would then have become the ring system we see today.

In Fig. 8.3 is a graphic showing Saturn's rings. As you can see, the naming of the rings is slightly unusual; they were named as they were discovered. Thanks to *Voyagers 1* and *2,* and more recently the highly successful Cassini mission, we now know a great deal about the rings. We shall give a brief survey of them here, starting with the innermost and working outwards.

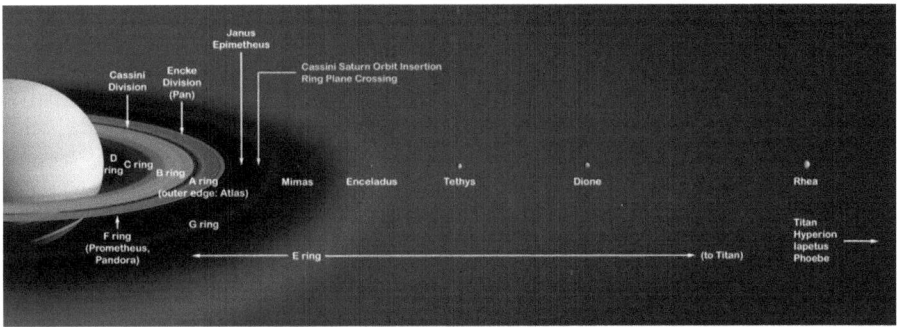

Fig. 8.3 Graphic showing the layout of Saturn's rings, and the satellites within them (Courtesy of NASA)

D-Ring

The innermost ring was discovered by *Voyager 1*. The ring is not a proper ring in that its edges are not well defined. The D-ring is not visible in amateur telescopes.

C-Ring

Also known as the Crepe ring due its semi-transparent nature, it was discovered by William and George Bond in 1850 (although William Dawes and John Galle also observed it independently from the Bonds). The C-ring is rather dark, but it can be seen in a 6 in. (150 mm) reflector in good conditions. When the rings are edgewise, or almost edgewise, the ring becomes invisible.

B-Ring

This is both the brightest and the largest of the rings and is visible even in small telescopes when the planet is not edgewise on. The B-ring does vary in brightness, and larger telescopes (8 in. (203 mm) or more) will reveal the darker inner B2-ring, and a outer brighter B1-ring.

The Voyager spacecraft recorded interesting spoke-like features that came and went with the planet's rotation. They have been seen by amateurs over the years with telescopes apertures of 8 in. (203 mm) or larger, in good conditions.

Cassini Division

This is an apparent gap (although Voyager showed it to contain ringlets and particles) between the B-ring and the A-ring. It is named after Giovanni Cassini, who first recorded it in 1675. When the rings are wide open, the division can be seen in a 5 in. (127 mm) telescope, and in larger telescopes you will be able to trace the division all the way around. When the rings are narrow, it can be hard to see much of it even in large amateur telescopes. Normally the Cassini division appears jet black, but in periods of poor seeing it can look lighter and grayish.

A-Ring

The second most brightest ring, normally yellowish in color, but it has been observed to show in browns and grayish tones. Visible in good small to medium telescopes, larger telescopes (8 in. (203 mm)) show the ring to have slightly darker edges. Very large telescopes (12 in. (304.8 mm) or more) will reveal the Encke division under good conditions.

F-Ring

This is a rather thin and somewhat irregular ring well beyond the reach of Earth-based telescopes. It was first sighted by the spacecraft *Pioneer 11*. The Cassini spacecraft revealed that this remarkable structure is made of a 'core' ring with a spiral of material surrounding it.

Janus-Epimetheus Ring

Just beyond the F-ring, this small ring was probably formed by material originating from its shepherd moons Epimetheus and Janus. The ring was discovered by the Cassini spacecraft team in 2006.

G-Ring

This is a thin faint ring and contains a bright arc near the inner edge. The satellite Mimas orbits inside the ring.

Fig. 8.4 Drawing by David Gray showing extensions to the ring. The observation was made on 1980, March 4, at 0240UT with 415 mm DK at ×270. The extensions go out as far as Enceladus, suggesting it is the E-ring that is visible (Courtesy David Gray)

Pallene Ring

This is a faint ring formed by the dust emanating from meteorite impacts on the surface of the satellite Pallene. It was discovered by the Cassini imaging team.

E-Ring

This is the outermost ring; it is faint, and its boundaries are rather ill defined. The ring starts at the orbit of Mimas and then tails off somewhere close to the orbit of Rhea. It contains the orbit of the highly cryo-volcanic moon, Enceladus. Indeed, it is now supposed that much of the material of the E-ring has emanated from Enceladus.

In the journal of the British Astronomical Association (Vol. 118, 2008, p. 230) British astronomer and historian Richard Baum wrote an excellent paper documenting occasional observations of Saturn's E-ring by some observers during the last 200 years or so, when the rings were edgewise. The most recent example is the drawing made by David Gray on the 1980, March 4, (see Fig. 8.4). We know that Enceladus is incredibly volcanic, and it seems reasonable that if the satellite were to undergo a major outbreak and deposit a great deal of ice in the ring at the time of the ring plane crossing, then the E-ring *may* temporarily become visible. It is certainly worth watching for extensions to the rings during edgewise presentations.

Phoebe Ring

This is a rather tenuous ring in the orbit of the satellite Phoebe. The ring was discovered by the NASA Spitzer infrared telescope. The ring is inclined to the ring plane by an angle of some 27°, as is the orbit of Phoebe, clearly implying the material came from the satellite.

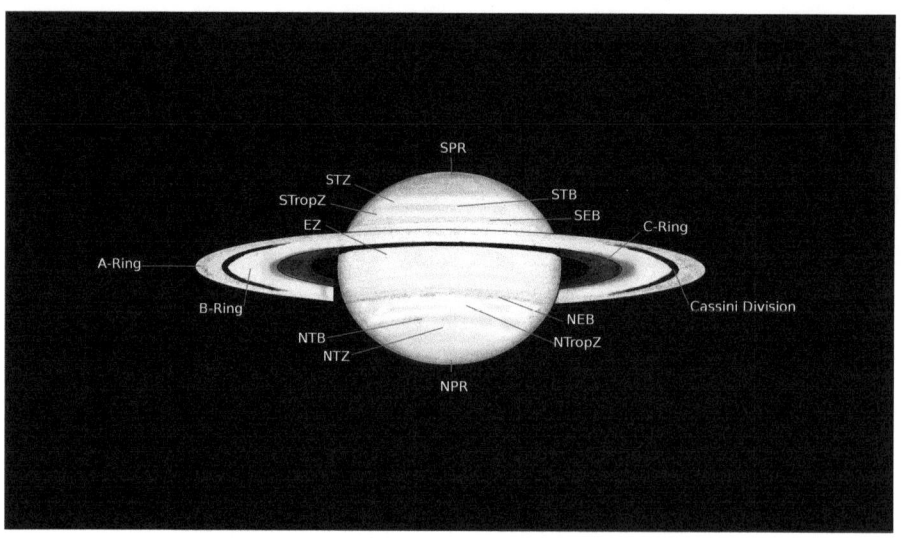

Fig. 8.5 The features of Saturn using BAA notation and naming

Features on the Planet

Let us now turn our attention to the various features visible on the planet though amateur telescopes. In Fig. 8.5 are the names and nomenclature as adopted by the BAA and most other organizations. Everything within the EZ falls into System I longitude; everything else is in System II. NASA and other organizations tend to use System III (the internal rotation of Saturn) as a longitude measurement, and it is common now to include all three values of longitude on observations of the planet.

We shall now give a brief overview of the various features, starting the south and working northwards.

South Polar Region (SPR)

A grayish yellow region at the south pole of the planet. Sometimes a darker southern polar cap can observed here, though not always. Occasionally, the region contains a bluish or greenish coloration.

South South Temperate Zone (SSTZ)

A light yellowish zone just north of the SPR. This feature is not that easy to observe when present and probably requires an 8 in. (203 mm) 'scope to see it convincingly. The occasional white oval has been observed in this region albeit rather infrequently.

South South Temperate Belt (SSTB)

When present, this takes the form of a vague grayish belt. It requires a telescope of 8 in. (203 mm) or larger to observe it.

South Temperate Zone (STZ)

This is a rather bright conspicuous zone, usually creamy yellow in color. This zone does seem to vary in intensity during the course of an apparition, and at times it can rival the EZ, while at other times it may be no fainter than the STropZ to the north. The zone can be seen in a 4.5 in. (114 mm) reflector. Occasional white ovals have been observed here.

South Temperate Belt (STB)

A grayish brown belt that can be quite conspicuous, it can be observed in a 4.5 in. (114 mm) reflector, and larger telescopes show occasional slight irregularities along its boundaries or within it.

South Tropical Zone (STropZ)

This is a bright zone that can be seen in 3 in. (76.2 mm) 'scopes under good conditions. The zone is normally a yellowish cream color but can take on subtle pastel shades of green and blue. Spots and white ovals have been reported in this region.

South Equatorial Belt (SEB)

Normally the SEB is one of the most conspicuous belts on the planet, and often can be seen in a good 3 in. (76.2 mm) refractor. In 6 in. (150 mm) telescopes or larger the belt frequently appears as double, with a south component (SEBs) and north component (SEBn) separated by a zone (SEBz). The belt normally appears a grayish-brown color. Often the SEBn is the darker and broader of the two components, though this is not always the case. Slight irregularities along the southern edge of the SEB(s) have been observed as well as darker sections and spots within both components of the belt. Over the years, faint extensions have been observed extending down from the northern edge of the SEBn into the brighter EZ.

Shadow of Rings on the Globe (ShRG)

The rings cast a dark and prominent shadow onto the disk of the planet. Before opposition the ring shadow is located north of the rings in the EZ; at opposition it practically disappears and afterwards appears north of the rings. When viewed far from opposition it can be quite prominent as a jet black line. Larger telescopes will show curves towards the edges of the planet.

Equatorial Zone (EZ)

A broad bright zone around the equator of the planet. Very often the zone is split into two with a southern component (EZs) and a northern component (EZn). Sometimes a faint Equatorial Belt (EB) may separate the two regions but this is not always so, and when present, the belt sometimes is not present at all longitudes. Over the last few apparitions it seems that the EZ(s) may have been the darker of the two parts.

The EZ is home to a truly remarkable phenomena on Saturn – a Great White Spot event. It seems that (approximately) every 30 years, the EZn of Saturn produces an enormous, spectacular white oval that is much more brilliant than the surrounding EZ. As we shall see in the historical accounts, W. T. Hay recorded such a storm in 1933 with his 6 in. (150 mm) refractor, and similar ones were observed in 1960 and 1990. Generally, the oval will be brilliant white and may span the height of the EZ. Eventually, the powerful winds in the EZ will disrupt the oval and spread the material throughout the whole of the zone. The EZ will become much brighter as a result until finally it returns to its usual appearance and all is quiet once more.

The white ovals themselves are thought to be caused by a large upwelling of ammonia from deep down in the atmosphere. The ammonia is carried up to the top cloud deck, where it freezes out into brilliant white crystals, making the oval very distinctive. The exact mechanism (and reason) this occurs is somewhat elusive, but it seems that the appearance of these ovals is connected with the summer solstice of the northern hemisphere. Saturn is a very dynamic world, and there may well be other explanations for the appearance of GWS.

North Equatorial Belt (NEB)

Another very prominent belt, often brownish-orange in color (although it can also have a distinct bluish tone). Often, like its southern counterpart, the NEB is double with a southern component (NEBs) and a northern component (NEBn) separated by a graying zone (NEBz). Usually the NEB components are darker than the SEB ones, and quite often in larger telescopes (8 in. (203 mm) or more) the edges of the belt appear to undulate rather than be completely uniform. Occasional darker sections within the belt and filaments connecting the two components have been observed fleetingly in the past.

North Tropical Zone (NTropZ)

A bright yellowish zone just north of the NEB, although like the neighboring NEB, the region can take on a bluish appearance. Normally the zone is quiet, but in 2010, the NTropZ was host to one of the most powerful storms ever seen on Saturn. A brilliant white spot erupted and spread throughout the zone (see section later on storms), disrupting the northern edge of the NEB and the NTB. Certainly it was one of the most dramatic events ever observed on the planet!

North Temperate Belt (NTB)

This is a thin grayish-tan feature (although it had a distinct orange-red component of color in 2012) that can be hard to discern at times and is not always visible. The belt has demonstrated some variation in intensity during the course of an apparition.

North Temperate Zone (NTZ)

This is a bright yellowish zone, although subtle hues of blue or green can be present. Occasional white storms have been reported here.

North North Temperate Belt (NNTB)

This rather vague thin grayish belt is sometimes visible with large telescopes (8 in. (203 mm) or larger).

Fig. 8.6 A NASA Cassini image showing the northern polar storm taken on 2012, November 27th, Cassini was at a distance of about 224,618 miles (361,488 km) from the planet. The wonderfully complex structures of the storm are nicely shown

North North Temperate Zone (NNTZ)

This is an occasional brighter zone just south of the NPR visible in larger telescopes.

North Polar Region (NPR)

A dark grayish region at the northern pole of the planet, this area occasionally sports a darker north polar cap (NPC) that might also be seen in larger telescopes. Saturn's north pole is home to a stunning phenomena known as the North Polar Storm. This vast hexagonal shaped feature was first imaged by the Spacecraft and has been recently imaged in magnificent detail by the CASSINI spacecraft. It is thought that the hexagonal shape has arisen due to the meeting of six jet streams that have created standing waves. Whatever the reason for its existed it must be one of the most remarkable structures in the solar system. During the 2012–13 apparition a number of images were able to capture the hexagonal structure, and visual observer David Gray was able to detect its structure with his 415 mm DK telescope.

Shadow of the Globe on the Rings (ShGR)

The globe of Saturn casts a large shadow on the rings, and this is visible from Earth. Even a small telescope will show it when the shadow is prominent. Before opposition the shadow lies on the preceding side of the ansae (another name for the ring system); at opposition it vanishes and then returns to the following side after opposition. The shadow is quite large when the planet is well before or well past opposition.

Satellites

Saturn has a vast number of satellites. The current count is approximately 200, although it's likely that more will be discovered at some point in the future. No less than seven of the satellites (Mimas, Enceladus, Tethys, Dione, Rhea, Titan and Iapetus) are visible in a 4-in. telescope. As the Cassini mission has shown us, the moons of Saturn are just as geologically active as those of Jupiter, but with cryo-volcanism being the main geological activity. The other satellites are dead, inactive worlds, but because it is worth having some familiarity with them, we shall give a brief survey of the seven main classical satellites here. We start with the inner ones.

Fig. 8.7 A montage of NASA images of the seven main moons of Saturn (Courtesy of NASA)

Mimas

With an apparent magnitude of about +12.8, this satellite can be glimpsed in a 4 in. (101.6 mm) telescope under good conditions. It was discovered by Sir William Herschel on September 17, 1781, using his vast 40-ft telescope. Mimas is a small rocky world whose surface is dominated by a large impact crater called Herschel. (In photographs with Herschel showing, Mimas always looks very much like the Death Star out of *Star Wars!*) Herschel Crater is about a third the diameter of Mimas itself and was clearly formed by an impact with a large body. It is a wonder the satellite was not destroyed.

Enceladus

This may be the most fascinating moon in the Solar System! Enceladus shines at a magnitude of +11.8 and so it is well within reach of a 4 in. (101.6 mm) telescope. Enceladus was discovered by Herschel in 1789, and the Voyager mission showed it to be a highly reflective icy world. We had to wait for the arrival of the Cassini spacecraft to reveal the full startling truth about this tiny moon. Enceladus is probably one of the most geologically active satellites in the Solar System, quite remarkable given its small size. The active geology takes place in the form of cryo-volcanism (cryo-volcanos erupt with ices and volatiles rather than lava). Cassini flew past this satellite several times and recorded enormous plumes of water ice crystals erupting from the surface and out into space. Some of this falls back onto the surface as snow, but a lot of it contributes to Saturn's E-ring, which the orbit of the Moon lies within.

The fountains of Enceladus are one of the greatest surprises our Solar System has offered so far. One does not expect this much activity on such a tiny moon. There are a number of explanations for the plumes, but the most likely origin of the jets is a small lake or sea under the surface of the moon kept warm by tidal heating.

Tethys

At magnitude +10.2 this is a rather bright satellite and should be visible in a 3 in. (76.2 mm) 'scope under good conditions. It is another classic satellite discovered by Giovanni Cassini in 1684. Unlike the lively Enceladus, Tethys does not appear to be active. The surface of Tethys is covered with impact craters, the largest of which has been named Odysseus. Tethys also has a vast trench that nearly stretches around the whole surface (nearly 270° of the surface). The trench is called Ithaca Chasma. There is still much debate as to whether some form of liquid water exists deep down under the surface of Tethys.

Dione

Like Tethys, it was discovered by Cassini 1684. With an apparent magnitude of +10.4 it should be visible in a 3 in. (76.2 mm) telescope, and certainly a 4 in. (101.6 mm) will reveal it. The icy surface of Dione seems to be a mixture of old heavily cratered regions and more recent smooth planes. The moon is perhaps most famous for its vast icy cliffs. First imaged by the Voyager spacecraft in 1981, its exact nature could not be determined. Cassini provided many hi-res images of the satellite and showed the features to be vast ice cliffs, probably formed by tectonic fractures.

Rhea

After Titan, Rhea is the next largest moon of the Saturnian system. Interestingly, Rhea's leading surface is bright and heavily cratered, while the other side (the trailing hemisphere) seems to be somewhat darker and contains a fair number of craters. It was thought that the moon might have a ring system; this was hypothesized to explain the depletion of electrons in Saturn's magnetosphere in the vicinity of Rhea. However searches have revealed no rings, and so the electron loss is still a mystery at the time of writing. Models of the interior predict temperatures warm enough to support a liquid ocean.

Titan

This is the largest of Saturn's satellites and the only satellite in the Solar System to have a substantial atmosphere that gives a surface pressure of 1.5 atm. Titan was first observed by Christiaan Huygens in 1655, and at magnitude +8.4 it is visible in 10×50 binoculars. The atmosphere of Titan was first shown to exist by Gerard Kuiper in 1944. It completely covers the surface, and even the Voyager spacecraft were unable to penetrate the thick atmospheric haze. On 2005, January 4, the Huygens spacecraft touched down safely on the surface of Titan. The spacecraft returned pictures to Earth showing what looked like a shoreline, dark rivers and hills. Subsequent flybys have revealed lakes composed of methane; the hydrocarbon lakes make Titan the only other place (after Earth) in the Solar System to have standing liquids on its surface.

In large telescopes (8 in. (203 mm) or more), Titan presents a tiny orange disk. Recent observations by David Gray, Alan W. Heath and this author seem to suggest that the satellite may have become a little redder in color (although at the time of writing, Saturn is low in the northern hemisphere, and this may easily account for the color change). Certainly the satellite is worth watching to see if there is any an interesting variation in magnitude or any slight color change.

Iapetus

We finally come to the last of the large satellites, Iapetus. This is an interesting world for a number of reasons. First its orbit is inclined to the ring plane by some 14.7°. Secondly, the planet has a definite two-toned surface. The Cassini Regio area on much of the leading hemisphere is incredibly dark and appears to be coated in a black substance. In contrast the trailing hemisphere is white and very bright. Both surfaces are heavily cratered. The two-tone effect is so pronounced that it has an influence on the apparent magnitude of Iapetus. When on the western side of Saturn (the left hand side in a normal telescopic inverted view with north at the bottom and no star diagonal used), the moon has an apparent magnitude of +10.2. However when it moves to the eastern side it fades by two magnitudes, down to +11.8. This means that if you observe Saturn with a 2 in. (50.8 mm) 'scope, you will see Iapetus when on the western side, but it will become invisible on the eastern side. One explanation for the dark material is that it comes from the Phoebe ring, which touches the surface of the moon.

In 2004, a vast ridge running down the center of Cassini Regio was discovered by the Cassini team; the reason for its existence is unknown at this time.

Historical Accounts

Saturn is one of the five bright classical planets that would have been visible to the earliest ancestors of the human race, and its steady movement through the heavens was noted by many classical civilizations. The invention of the telescope revealed to human eyes something that had never been seen before – a planet circled by a magnificent ring system.

Galileo was the first of the telescopic explorers, and he turned his telescope towards Saturn sometime in July 1610. Although his telescope was less powerful than most viewfinders today, he was, nonetheless, able to discern something strange about the shape of Saturn. Over the years, Galileo made a number of drawings of Saturn, one showing Saturn with a large satellite either side, one with Saturn alone (and no rings) and other drawings showing the attendants once more. Galileo was unable to conclude that he was seeing a ring system; his telescope did not have the resolving power. He could not have known that the reason for the changing view was due to the changing aspect of Saturn's tilt to Earth as it journeyed around the Sun.

Galileo was not alone. The French astronomer and philosopher Pierre Gassendi, who had made the first-ever recorded observation of the transit of Mercury, observed Saturn. His drawing of the planet appears to show a ring, but like Galileo he was unable to reach the correct conclusion about what he had observed. The true nature of this puzzling view of Saturn was finally discovered by the Dutch astronomer and Renaissance man, Christiaan Huygens.

Huygens was a careful and skilled observer. He discovered Saturn's largest satellite, Titan, in 1655. He correctly realized that Saturn was surrounded by a ring and went on to correctly explain the changing aspect of Saturn's rings. All of this was published in his magnificent treatise "Systema Saturnium." Of course, not everyone accepted the ring hypothesis, and there were challenges to Huygens ideas. The alternative ideas proposed were quite wild, and thankfully Huygen's ideas of the Saturnian system held sway.

By the late 1660s other observers started to make their own observations of Saturn. In particular, Robert Hooke of England made a fine drawing of the planet on 1666, June 16. The drawing is not too dissimilar from a modern Saturn drawing, the rings correctly drawn and what looks like a hint of the Crepe Ring a number of years before it was discovered.

The Italian astronomer Giovanni Cassini (also known as Jean Dominique Cassini) had established a reputation for himself as an excellent observer when he was professor of astronomy at the University of Bologna – so much so in fact that King Louis XIV invited Cassini to supervise the new Paris Observatory in 1671. Cassini made good use of the telescope; in 1670 he was the first to observe the division between the A- and B-ring, now justly named the Cassini division, and in 1684 he discovered the satellites Dione and Tethys.

It seems there was something of a lull in observations until the mid eighteenth century, possibly because the planet does not display the striking markings of Mars or Jupiter. Charles Messier, who famously drew up his catalog of deep sky objects so people would not mistake them for comets, observed the planet in 1766 and was able to make out the main equatorial belts.

Perhaps the greatest contribution in the eighteenth century came from Sir William Herschel, the famous Hanoverian astronomer who discovered the planet Uranus. Even judged by our times, Herschel was a remarkable man, and made many important contributions to planetary astronomy, as we have seen in earlier chapters. Among his achievements, he discovered the satellites of Enceladus and Mimas in 1789 and made observations of the belts and atmosphere of Saturn. Interestingly, Herschel believed the rings of Saturn to be solid. His argument was quite persuasive; he noticed the shadow of the rings on the globe that could (he thought) only be made by a solid object. He also observed that Saturn was decidedly oblate and bulged near its center – again he attributed this to the gravitational pull of a solid ring.

Herschel also made a number intriguing observations that could be put down to poor conditions or the limits of optical components of the time. The first was the observation of a number of 'star-like' points on the rings that seemed to keep pace with the rotational period Saturn. He was quite unable to explain this but seemed convinced of their reality. Herschel made one other unusual observation of interest to us here. In 1805 he believed that the figure of Saturn had changed. Instead of gradual polar flattening, like that of Jupiter, he believed it had become sharper, and as a result the planet took on a slightly more oblong shape rather than the traditional ellipse. This view of Saturn persisted for Herschel into 1806 and was the subject of two of his papers.

Herschel did not just take his visual observations at face value, he tried micrometer measurements to determine the latitudes of the points of greatest curvature on the globe and examined the planet with different telescopes and eyepieces. Herschel was unable to explain the effect, but today it seems as though – as the famous Saturn observer A. F. O'D Alexander notes in his book *The Planet Saturn, A history of observation, theory and discovery* – Herschel had been "the victim of a very subtle optical trick by Saturn to which ring position, the polar shading and the planet's low altitude all probably contributed."

The next series of useful observations came from William C. Bond and his son George. The Bonds discovered the semi-translucent C-ring in 1850 (although it should be noted that William Dawes and Johann Galle observed it independently of the Bonds). The Bonds made an excellent series of observations from 1848, June 25, until 1849, January 19, using the 15 in. (381 mm) Merz refractor at Harvard. During this time, Saturn was edge-on to Earth, and the Bonds observed that the rings did not always appear as a single thread of light; at times it appeared to them as though the rings had a distinctive 'globule'-like appearance.

Well defined spots continued to appear on the planet from time to time, and many astronomers followed Hershel's lead in using them to derive a rotational period of Saturn. Professor Asaph Hall, while at the U. S. Naval Observatory in Washington, used the large 26 in. (660.4 mm) refractor (the largest refractor in the world at that time) there. He was able to use a bright white spot to determine a rotational period of 10 h 45 min.

Amateur astronomers continued to monitor the planet. A bright spot was seen in 1903 by A. S. Williams, but the next large event seen by amateurs was the Great White Spot of 1933. The credit of the discovery of this spot goes to Will Hay, the famous stage, radio and film comedian. In almost all of his films, Hay plays a bumbling professional who lurches from one comedic disaster to another, desperately trying to cover his incompetence. In real life however he was a very serious man, and an avid amateur astronomer. He set up his observatory and equipped it with a fine 6 in. (150 mm) refractor. On the 1933, August 3, at 2235UT, Hay was observing Saturn when he noticed a brilliant white spot in the EZ. He reported it to the BAA, who got the word out, and many astronomers were able to watch as the storm was dispersed along the EZ. A similar white spot was reported by Botham in 1960, and another one erupted in 1990.

Today amateur astronomers are still keeping up their vigil and continue to monitor Saturn. Although it is true that we have the excellent Cassini mission out at Saturn, modern amateurs play an important part in monitoring the whole of the Saturnian system rather than just one part of it. As we have seen here (albeit briefly), today's observers have a long and rich heritage to build on.

Observing Saturn

A small telescope of 2 in. (50.8 mm) will show you the rings and moons, and perhaps an equatorial belt if they are particularly prominent. For serious study, however, a telescope with an aperture of 6 in. (150 mm) is required. Saturn tolerates magnification reasonably well, and on a night with average conditions, powers of

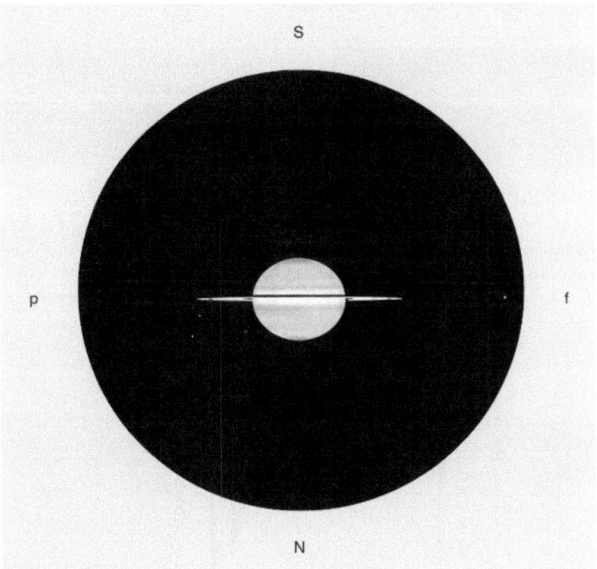

Fig. 8.8 Drawing of an almost edge-on Saturn by Carlos Hernandez on 2010, June 10, 0145UT, with a 228.6 Maksutov-Cassegrain at ×295 & ×388 (Courtesy Carlos Hernandez)

between ×167 and ×250 should be more than adequate to show the subtle features in the belts, zones and rings. On good nights, using a power of ×400 with a 8 in. (203 mm) reflector can result in some truly superb views of the planet!

As remarked earlier, the features of Saturn are rather muted, so it is best to resist using a really high power unless the conditions are good, and the image of the planet is bright (Fig. 8.8).

Advanced Projects

Disk Drawings

In order to execute a disk drawing of Saturn, a blank of correct dimensions and the correct tilt is essential. It is best not to draw the planet's outline directly from the telescope, as to do so requires at least five ellipses to be drawn accurately! The Saturn section website of the BAA has Saturn blanks that can be downloaded that are of the correct size, orientation and scale. The *Sky at Night* magazine also produces them.

Before you print off a blank to use at the telescope, you will need to know the correct value of Saturn's tilt. *WINJUPOS* will tell you the current value of Saturn's

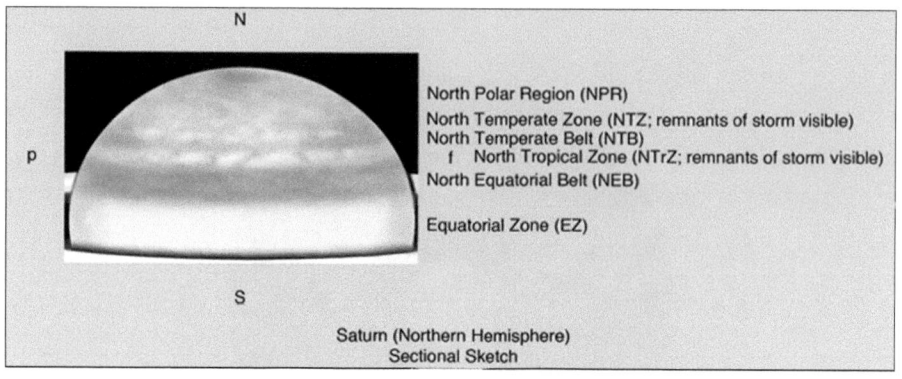

Fig. 8.9 Sectional drawing of Saturn by Carlos Hernandez on 2011, April 20, at 0230UT with a228.6 mm Maksutov-Cassegrain ×388. Remnants of the 2010 storm in the NTropZ and NTZ are visible (Courtesy Carlos Hernandez)

tilt; simply input the date and in the Ephemerides, the "declination of the earth" is the value of B. You then need to use a blank that has this B value.

You should spend about 30 minutes looking at Saturn before you begin to draw. Try to examine each belt and zone carefully, so you have a good idea what is present in your mind when you begin to make a drawing.

Like Jupiter, Saturn has a rapid rotation, so it is best to complete your drawing in under 12 minutes. When you're ready to go, start by drawing in the main features (You can start with the equatorial belts). Also include any obvious features on the rings (Cassini Division, C-Ring if seen, etc.). This should take about 6 minutes. Finally, spend the final 6 minutes putting in any fine details in the belts, zones (dark condensations, for example) and rings. When you have completed that, you can add on any shadows that are present and record the time (in UT) and other observational details.

Sectional Drawings

It is not uncommon for Saturn to produce small short-lived storms, and a sectional drawing is the best way to record this. It might be the case that the storm is best seen in a particular filter, so you should experiment both with magnification and filters to get the best view. When you have done this, you can draw the feature and any of the surrounding environment the object seems to be affecting. Make sure you record the time and observing details, along with any remarks about color and intensity alongside the drawing (Fig. 8.9).

Fig. 8.10 Drawing of the 2010 storm by the author on 2011, May 12, at 2322UT, with an 8-in. Newtonian at ×167 with a *blue* W#80A filter. The storm was particularly prominent in this light

Filter Work

Both Saturn and its ring system are good subjects for filter work. You will find the following filters helpful:

- W#8 (yellow): This is a useful filter for enhancing the contrasts between the belts and zones.
- W#25A (red): This filter increases bluish colored features, it is also helps to steady the seeing the view when seeing is bad, since longer wavelengths of light are not as affected by poor seeing as the shorter (bluer) ones.
- W#80A (light blue): This filter helps emphasize reddish-orange features such as the main equatorial belts. The storm of 2010 was particularly prominent in this filter.
- W#30 (light magenta): This filter allows the passage of both red and blue wavelengths but stops the transmission of green. The W#30 is useful for increasing the contrast of low contrast markings.
- W#47 (violet): This filter is rather good at increasing the contrast in the rings with larger telescopes of apertures of 8 in. (203 mm) or more.

If there is a specific feature on the globe or rings that is emphasized in a particular filter, it is best to make a drawing of it in that filter. Figure 8.10 is a drawing of the powerful 2010 storm that was most dramatic seen using a W#80A filter.

Making Intensity Estimates

The various belts and zones (and indeed the rings) of Saturn do demonstrate fluctuations in brightness over the course of an apparition. The BAA uses a scale of 0 (extremely bright) to 10 (black sky), and here we should add a word of caution. In the past, it has been taken as standard that the B-ring is of value 1. However, fluctuations in the brightness of this ring have been observed over the years, so it's best to record the intensity of this ring in the same way as other features.

As with Jupiter, make a list (from south to north) of all of the features that can be seen on the disk, and a similar one for the rings (from A-ring, through to C-ring). Once you have done this assign each feature a value of intensity. It is interesting to do this with a number of filters (make estimates in a red W#25A and a blue W#80A) to see how any changes vary in different wavelengths of light.

The Bi-Colored Aspect of the Rings

This is an interesting phenomena that has been seen by a number of observers over the years. Occasionally, it seems that the A-ring (and sometimes the B-ring) appear to be brighter in a blue filter (like a W#80A) than in a red one (like a W#25A). Very often the effect is limited to the rings on one particular side of Saturn rather than appearing on both the east and west ansae. It is not at all certain whether the effect is caused by some dynamical interaction of the ring particles or is an effect of our own atmosphere. Only repeat observations will help solve the mystery, so it is worth checking the rings with a red and a blue filter to see if the effect is present.

Storms

Saturn's atmosphere lacks the striking long-term storms that are present on Jupiter. It is not uncommon during the course of an apparition for the planet to produce white ovals in various belts or zones, but these are by and large quite rare and usually short lived. Sometimes, however, Saturn's atmosphere is capable of generating some truly epic storms. These storms may change the appearance of a whole zone and last for many months after their initial outbreak. We shall look at three types of storms here:

GREAT WHITE SPOTS (GWS) As mentioned in the EZ section above, these seem to be seasonal storms that occur approximately every 30 years. They take the form of a vast brilliant white oval in the EZ and are quite unmistakable. If you should see one, report it at once to the Saturn section director of the BAA. You should also try and estimate when the center of it is on the CM. This will allow you to determine the CM1, CM2 and CM3 longitudes, which will be vital in tracking it. Make sure you include all your observational details when you report it.

You will find that over the course of a few weeks, a GWS will start to lose its well defined boundary and eventually will be dispersed throughout the whole EZ. You should monitor the region closely to see what changes are occurring, both in IL and with filters. Intensity estimates will also help provide details on how the zone is brightening and how long it takes for the material to circulate through the whole of the zone. Sectional drawings may be more appropriate if subtle details are present at high magnifications or with certain filters.

Fig. 8.11 Beautiful image from the Cassini spacecraft showing the storm of 2010 raging in the atmosphere of the planet (Courtesy of NASA)

SMALLER REGIONAL STORMS As the storm of 2010 showed, not every dramatic storm on Saturn takes the form of an EZ oval. On 2010, December 5, the Cassini spacecraft detected some very powerful lightning in the planet's atmosphere and produced up to 10 lightning flashes a second. Soon after a brilliant white spot appeared that dramatically evolved a brilliant 'tail' as material was dispersed from the bright core by the powerful winds in the NTropZ (Fig. 8.11). After a few months the material had stretched through much of the NTropZ and had disrupted the NTB. The storm dominated Saturn for the 2010–2011 apparition, and there even appeared to be slight remnants of it in 2012. This type of storm seems to be very rare, but obviously no one knows when another will erupt. Monitoring for this type of storm is therefore another useful task the visual observer can do. As in the case of GWS, if you do suspect something, report it and try to obtain a transit time so you can establish its longitude.

SMALL WHITE OVALS Observers with telescopes of 8 in. (203 mm) or larger should keep an eye out for small white ovals that can appear anywhere. Normally they are quite short-lived and may vanish after one rotation. One way to determine if your suspected spot is real is to check that it moves with Saturn's rotation. You might find a W#80A filter helps to enhance it, and of course you should obtain a longitude estimate so other people can find them.

Saturn is a thoroughly fascinating world to study. Very often it takes years of patient observing before anything striking happens, but when Saturn does put on a show, it's very often powerful and dramatic so, be ready for it!

Chapter 9

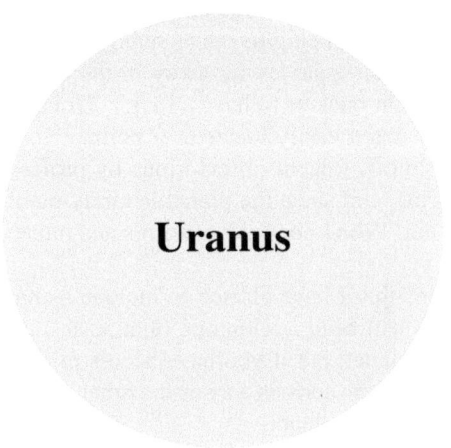

Uranus

Distance from the Sun: 3,003,623,00 km (max.), 3,303,623,00 km (min.)
Diameter: 51,118 km
Sidereal Period: 84.01 years
Synodic Period: 369.66 days
Rotational Period: System I (Magnetic field): 17 h 14.4 min
Mean Surface Temperature: −214°C
Apparent Diameter: 4.1″ (Max.), 3.1″ (Min.)
Albedo: 0.51
Brightest Apparent Magnitude: +5.6
Satellites: 27 (known)

We have reached the cold outer wastes of the Solar System, and the last planet of the Solar System that the visual observer can usefully study. For much of the last century up until today, Uranus has been neglected. It has languished in the dark, lonely parts of the Solar System relatively unobserved. There have been good reasons for this. The planet is low down for northern hemisphere observers, and that combined with its tiny disk size means that only observers with large telescopes could observe it. Large telescopes were scarce in the mid to late twentieth century,

P.G. Abel, *Visual Lunar and Planetary Astronomy*, The Patrick Moore
Practical Astronomy Series, DOI 10.1007/978-1-4614-7019-9_9,
© Springer Science+Business Media New York 2013

and it is only in the last few decades that enormous Dobsonians have become available.

The increase in the availability of large telescopes has seen many people take up deep-sky astronomy, the game being to see as many of the Messier objects as possible, or to find elusive faint galaxies far away in the cosmos. It seems very few people have used their light buckets to have a look at Uranus. Why is that? Perhaps it is because the planet is not easily located, or perhaps its imposed reputation of being bland puts people off. Recent observations by professional telescopes have shown clouds and storms, and since the planet is rarely monitored it is hard to say whether it is bland or not! Who knows what storms and interesting phenomena may have been missed.

It is for these reasons that I have chosen to include a chapter on Uranus in this book. It is true that you will need a telescope of at least 8 in. (203 mm) or more if you intend to study the planet, but if you have access to a large instrument, I urge you to treat Uranus in the same serious way as the other planets and establish a clear observational program for this planet.

Orbit and Apparitions

Uranus moves around the Sun very slowly, taking some 84 years to complete one circuit. As a result it spends quite a long time in any given constellation. For northern hemisphere observers things are rapidly improving, as the planet is now above the celestial equator and is climbing out of the southern depths. At the time of writing (September 2012) it is to be found among the stars of Pisces. Uranus comes to opposition every year and technically reaches naked-eye viability. The magnitude and disk size do not vary that much, and the slow movement of the planet in the sky means that once you have tracked it down, you should have little trouble finding it again the next night. Even in viewfinders the planet is obvious, shining as it does with a distinctive emerald hue among the background stars of the galaxy.

Uranus as a Planet

Uranus and Neptune are a different class of world from the other planets in the Solar System; they fall into the category of ice giants. Figure 9.1 shows the current theoretical model for the interior of Uranus.

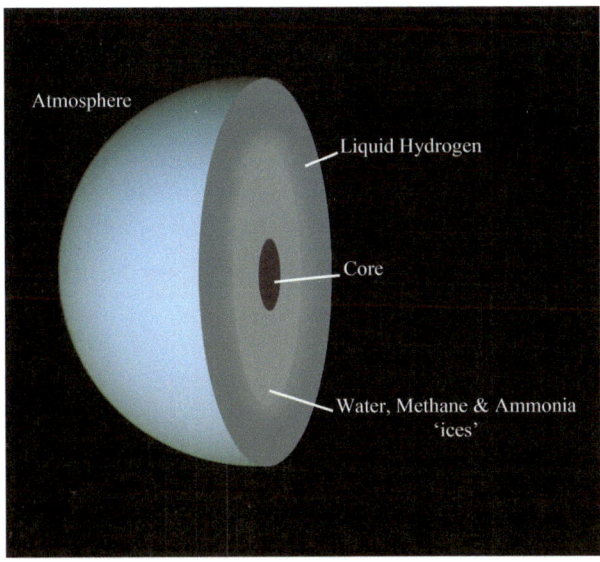

Fig. 9.1 Proposed interior of Uranus

At the top we have a thick atmosphere composed largely of molecular hydrogen, helium and methane. The atmosphere is a distinctive greenish hue that makes it quite unmistakable even at low telescopic powers. Underneath the atmosphere we come to a deep layer of liquid hydrogen. Beneath this we come to another large region, this time composed of ices – mainly water, methane and ammonia ice. Finally at the center of the planet, it is assumed that there is a rocky-silicate core. Unlike the other planets, Uranus seems to have virtually no internal heat, so there is some question as to whether the planet does have a well-differentiated core.

Another feature unique to Uranus is its pronounced axial tilt of some 97.9° – much more than a right angle! Quite why this should be the case remains obscure at the moment. It could be that Uranus had a collision with a large impactor in the distant past, and this knocked the planet on to its side. Such an impact may be the reason why the planet lacks a strong internal heat source. The axial tilt of Uranus has made the seasons on the planet really rather alien. For 42 years one pole of the planet will be in permanent sunlight, while the other will be in constant darkness. At the time of equinox, the equator receives full sunlight, producing day and night in the same fashion as the other planets.

The magnetic field of Uranus presents us with yet another mystery. It was *Voyager 2* that uncovered this peculiarity when it reached the planet in January 1986. In Fig. 9.2 we have the structure of Earth's magnetic field (a) and that of Uranus (b). The magnetic field of Uranus is inclined to the planet's axis of rotation by an angle of 29°; moreover the magnetic field does not pass through the center of

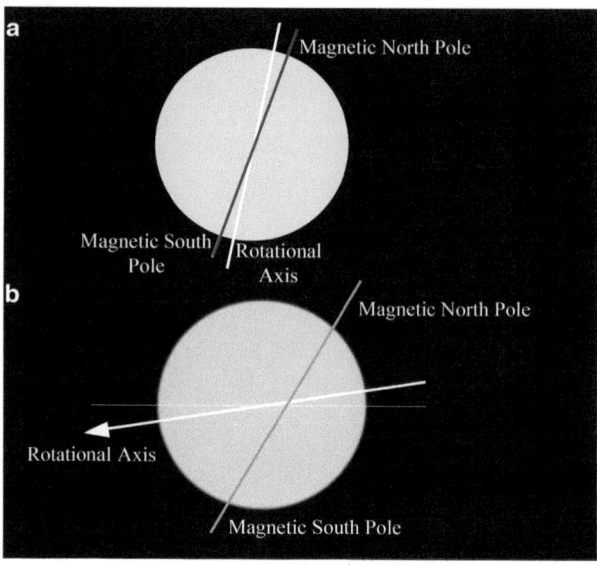

Fig. 9.2 (**a**) The magnetic field of Earth. (**b**) The magnetic field of Uranus

the planet. Rather, it is off to one side. This unusual location of the magnetic field means that the field strength can vary considerably on the "surface." It is interesting to note that the magnetic field of Neptune is similarly offset. One explanation for this is the location of where the field is generated. In terrestrial planets it is generated in the outer molted cores. In gas giants it is generated by the swirling metallic hydrogen and helium oceans under the deep atmosphere. With ice giants, it could be that the magnetic field is generated much higher up in the planet, perhaps in the oceans of ammonia and water ice.

As yet, the only spacecraft to visit Uranus was *Voyager 2,* and it is high time the planet was surveyed again! The answers to just some of these mysteries would be quite illuminating.

Rings and Moons

Uranus has a system of rings that was discovered by astronomers using ground-based telescopes here on Earth. Of course the rings cannot be seen visually, but their fingerprints were seen on 1977, March 10. On this date Gordon Taylor of the BAA had predicted the occultation of the star SAO 158687 by Uranus, and astronomers were preparing to observe it. The occultation was only visible in the southern hemisphere, and so the Kuiper Airborne Observatory was used to observe the occultation. The astronomers on board – J. Elliott, E. Dunham and D. Mink – noticed that the star winked several times some 35 min before the occultation and then did the same afterwards (the effect that occurred after the occultation was also observed independently by J. Churms in South Africa). The only possible explanation was that Uranus had a ring system, and it was the rings that caused the star to wink on and off.

The rings of Uranus are nowhere near as spectacular as those of Saturn. A total of 13 rings are known to exist, and they were imaged by the *Voyager 2* spacecraft. They seemed to be made of largely of particles of water ice. Since then, they have been imaged by the Hubble Space Telescope, but it has to be admitted that we still know very little about them.

The moons of Uranus are interesting objects and add further weight to the argument that some sort of catastrophe occurred in the distant past. A total of 27 moons are known, but only five of them are substantial and visible with amateur telescopes. Here is a short review of them, starting with the innermost moon of the five.

Miranda

This is a small moon discovered by Gerard Kuiper in 1948. As with all of the Uranian moons, the only data and images we have from it come from *Voyager 2*. The images sent back show the surface to be a right mish mash of surface features; there is old cratered terrain but also broken terrain and vast canyons all crisscrossing the surface. One theory to account for the appearance of Miranda is that it was smashed to pieces and reformed several times. Certainly that would account for its startling landscape! Miranda shines at an apparent magnitude of +15.8, so you will probably need a 16 in. (406.4 mm) 'scope to see it. The moon is quite close to Uranus, so glare from the planet is also a problem.

Ariel

This satellite was discovered by William Lassell in 1851. At magnitude +14.4 it might be glimpsed in a 8 in. (203 mm) 'scope in a dark sky using averted vision and should be visible in a 10 in. (250 mm) 'scope. Alas *Voyager 2* was only able to survey the southern hemisphere, but what could be seen shows a cratered surface, scarps and canyons. Interestingly the surface of Ariel seems to be younger than the surfaces of the other main satellites, and this suggests that some resurfacing took place in the past.

Umbriel

This is another satellite discovered by Lasell in 1851. The moon appears to be the most heavily cratered of the five main moons, with some substantial craters reaching almost 210 km in diameter. Most interesting, Wunda Crater seems to have a ring of bright, highly reflective material that looks quite striking in the images taken by *Voyager 2*.

Titania

Discovered by William Herschel in 1781, this is the largest of the Uranian moons and at magnitude +13.9 is also the brightest (indeed you might be able to glimpse this moon with an 8 in. (203 mm) reflector). The surface contains many craters and a number of large faults and valleys. In particular, the Massina Chasma runs from the equator to the south pole.

Oberon

Discovered by William Herschel in 1787, Oberon is the second brightest satellite and shines at an apparent magnitude of +14.1. It may be glimpsed in an 8 in. (203 mm) 'scope with averted vision in dark skies; certainly a 10 in. (250 mm) 'scope should show it. It is the most heavily cratered of the moons.

Historical Observations

Uranus was discovered by Sir William Herschel on 1781, March 13, and has the distinction of being the first planet ever discovered by a human being with a telescope. In spite of this, however, Herschel was not the first person ever to gaze upon this distant ice giant (although he was the first to recognize it for what it was). It is surprising to learn just how many times it was recorded before anyone figured out its true nature. For example, the first Astronomer Royal, Sir John Flamsteed, observed it on no less than six occasions. He seems to have believed Uranus to be a star, since he gave it the designation 34 Tauri! The French astronomer Pierre Charles Le Monnier observed it 12 times. This has always been puzzling, since the planet shows an obvious disk even at ×80!

At the time of the discovery, Herschel himself was happily surveying the stars of the Milky way with his homemade 6 in. (150 mm) Newtonian reflector. Herschel was an excellent mirror grinder and telescope maker. He made telescopes for a number of his contemporaries. He was examining the stars in the constellation of

Gemini when he came across a star that looked larger than the rest. He increased magnification and found the disk of the unusual object increased in size (something no star will ever do).

Initially, Herschel believed he had discovered a comet and sent in his observations to the Astronomer Royal of the time, Nevil Maskelyne, and of course to the comet astronomer Messier in Paris. As word got about, more astronomers made their observations and data accumulated. In order to establish what exactly the new object was, its orbit and size would need to be calculated (a long-winded process without a computer!). The Russian astronomer Anders Lexell computed its orbit and found that the new object was orbiting the Sun in an almost circular orbit some 19 Au from the Sun, with an orbital period of between 82 and 83 years (not a bad estimate). The most important thing was the size. Lexell determined an apparent diameter of between 3 and 5 seconds of arc. Clearly the new object could not be a comet. It had to be a new planet, and a large one if the calculations were even only approximately correct.

The new planet had to be named, and Nevil Maskelyne asked Herschel to name it. The king of England at the time was King George III, and so Herschel decided to name the new planet Georgium Sidus (i.e., George's star). Alas, the name was not very popular outside of England. Some astronomers suggested the new planet should be called Herschel. The German astronomer Johann Bode suggested the name Uranus, and this was more in keeping with the names of the planets. In mythology, Saturn had been the father of Jupiter, and similarly Ouranus (the first ruler of Olympus) was the father of Saturn. Eventually Uranus became the established name.

Herschel continued to observe Uranus and would go on to discover two of its satellites. During his early work with the planet, Herschel believed he had observed a ring that he noted was "a little inclined to the red." The ring of Uranus is indeed brighter in red; however, it is generally agreed that the rings are much too faint to have been observed with Herschel's telescopes (unless, of course, the rings were brighter then and have since dimmed), and Herschel himself never mentions them in his later investigations of the planet.

William Lassell discovered two further moons, and in 1862 he observed what he believed to be a bright spot near the center of the planet's disk. Further spots and zones were recorded using a 9 in. (228.6 mm) reflector in 1870, but on 1873, January 16, Lord Rosse's reflector showed nothing on the disk at all. Since then it seems there have been times when belts, zones and spots have been present on the planet, and other times when the disk has been blank. In 1902 H. Deslandres showed that the planet had a retrograde rotation, and these observations were eventually confirmed in 1911 by P. Lowell and V. M. Slipher at the Lowell Observatory. R. Mecke managed to identify methane (CH_4) in the atmosphere of Uranus in 1933. In 1977 the rings were identified, but the last major survey of the Uranian system has come from *Voyager 2*.

Earthbound astronomers still take some interest in the planet. A Hubble image of the planet taken on 1988, August 8, showed what looks like bright spots or storms on the disk. In more recent times UKIRT and the Gemini north telescopes have recorded a bright EZ and belts, and other observations would seem to suggest that the planet is not as dead and inert as we once believed.

Observing Uranus

Uranus can be a difficult target to observe because it is a faint object that requires a lot of power to see clearly. An 8 in. (203 mm) telescope is probably the smallest possible telescope that can be used to study the planet. With an 8 in. (203 mm) reflector, you can use powers of ×312 and ×400, which may enable you to see some faint surface features (Fig. 9.3).

The first difficulty comes in locating the planet. To do this, you will need to locate a good finder chart showing the position of the planet. It is best to use setting circles – Simply input the RA and Dec. However, if you are not familiar with this, or your telescope doesn't have an equatorial mount, you should scan the area of the sky where the planet is located with binoculars. Even in a pair of 10×50s the green-ish color of the planet makes it stand out. You can then track it down from there. Uranus moves slowly along its orbit, so once you have located it, you shouldn't have too much difficulty in finding it again.

You will need clear and transparent conditions to observe the planet. Any hint of mist and fog and the planet's light will be substantially reduced. Uranus will require powers of ×300 (or ideally ×400 or more), and if there is any murk in the atmosphere

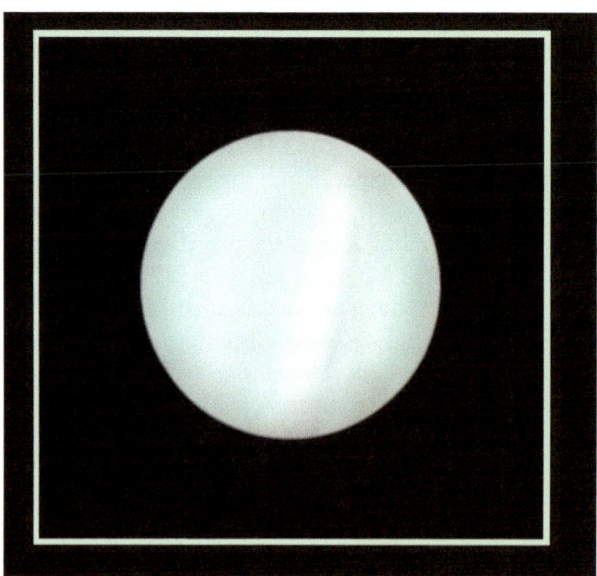

Fig. 9.3 Drawing made by the author of Uranus on 2011, November 17, at 2041UT with a 203 mm Newtonian reflector ×312

of Earth, the image will be too faint to be useful at these powers. Persistence is also the key when it comes to observing Uranus. There may be many months at a time when the planet is just a featureless disk. With so few observations it is hard to know how frequent surface features are seen on the planet.

Advanced Projects

Disk Drawings

These are the main contribution the visual observer can make. Uranus is not appreciably flattened, like the disks of Jupiter and Saturn, and so a circular blank 50 mm in diameter can be used to represent the disk of Uranus. It is best to spend about 10 minutes examining the surface to see what is present before you begin drawing. You will find that any surface markings on the planet are usually somewhat vague and ill defined, and averted vision is a good way of seeing them (Fig. 9.4). When you have recorded any features that are present, make sure you note down the time (in UT) along with the details of your telescope and atmospheric conditions.

You should also include the CM longitude (there is only one for Uranus), and this can be found on *WINJUPOS*. It is also a good idea to indicate north and 'following' on the disk so that anyone looking at your observation can establish the correct orientation (again you can get this from *WINJUPOS*).

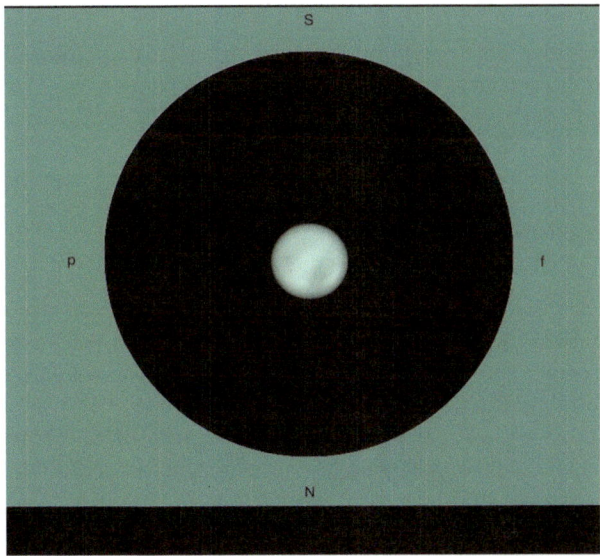

Fig. 9.4 Color drawing of Uranus by Carlos Hernandez on 2010, September 19, with a 228.6 mm Maksutov-Cassegrain telescope at ×310; a bright zone and darker region is visible (Courtesy Carlos Hernandez)

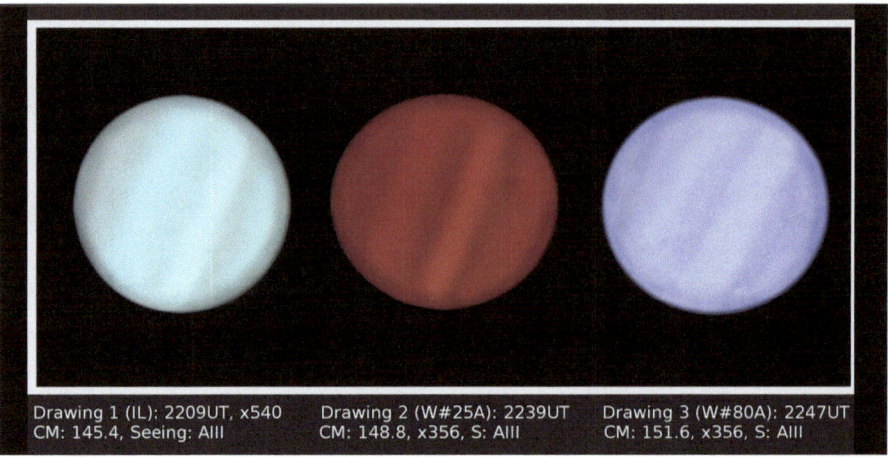

Drawing 1 (IL): 2209UT, x540   Drawing 2 (W#25A): 2239UT   Drawing 3 (W#80A): 2247UT
CM: 145.4, Seeing: AIII   CM: 148.8, x356, S: AIII   CM: 151.6, x356, S: AIII

Fig. 9.5 Filter drawings by the author made with the University of Leicester's 508 mm DK on 2012, October 6. Drawing 1 was made at 2209UT in IL, Drawing 2 at 2239UT in W#25A and Drawing 3 at 2247UT in W#80A

Filter Work

This requires a rather large aperture telescope, probably of aperture 15 in. (381 mm) or more. If you have access to a large telescope, try using a W#25A (red filter) to see how the limb darkening on the planet is affected (Fig. 9.5).

Intensity Estimates

If you have access to a 15 in. (381 mm) telescope (or larger) you may wish to assign intensity estimates to various parts of the disk, especially if you suspect there to be lighter zones/regions and belts to be present. You can use the BAA scale, which goes from 0 (bright white) to 10 (dark sky) (see Fig. 9.6). It would be nice to see some regular intensity estimates made over the course of a few apparitions, and I hope those people with very large telescopes consider undertaking this task.

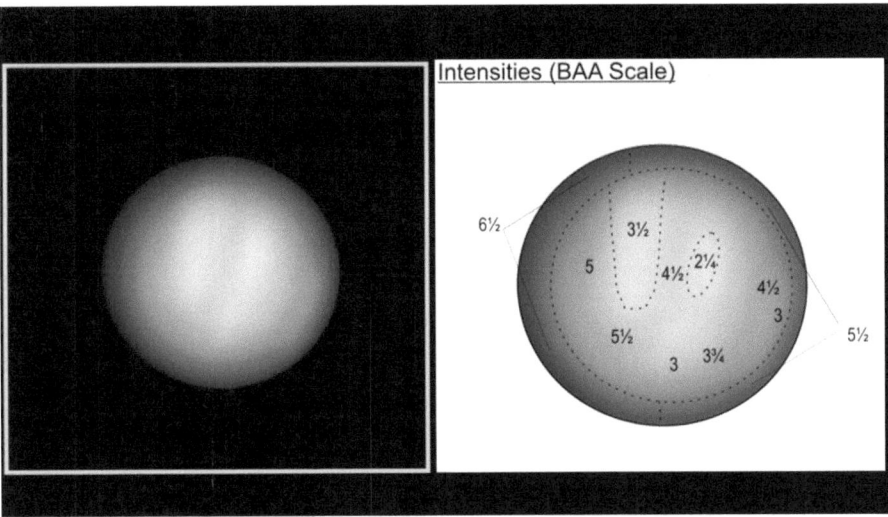

Fig. 9.6 A drawing of Uranus by David Gray on 2011, January 21, with his 415 mm DK at ×270 and ×535. Gray would also make some intensity estimates on the disk (Courtesy David Gray)

Magnitude estimates of the satellites would also be helpful in establishing if there are any fluctuations over the course of many apparitions.

Uranus has long been neglected. Who knows what storms and phenomena may have been missed. The planet is climbing higher for northern hemisphere observers, and many people have some access to large telescopes, so hopefully amateurs will consider observing the planet in the same serious way as we do the other planets. Who knows what surprises this distant ice giant has in store for us.

Appendix

Planetary Blanks

Contained here are planetary blanks for Mercury, Venus, Mars, Jupiter, Saturn and Uranus. These blanks can be scanned (or photocopied) and used at the telescope for drawing. They can also be used for your final neat drawings later.

In the case of Mercury, Venus, Mars and Uranus, use the circular planetary blank since, for all intents and purposes, these planets are fairly spherical. The Jupiter blank provided should be used for all Jupiter drawings since, as you can see, the planet is noticeably oblate.

Finally we come to the Saturn blanks. As Saturn goes around the Sun, its axial tilt means that we see different aspects of it. This tilt of the planet as viewed from Earth is called "B" and is measured in degrees. When Saturn's northern hemisphere is tilted B has a positive value; when Saturn's southern hemisphere is tilted we have a negative value. When the rings are edgewise on, B = 0. To find the value of B at the time of your observation, simply use *WINJUPOS* (free software). In the menu bar, go to Program, then Celestial Body and check Saturn. Next go to the Tools option in the menu bar and click on Ephemerides. This will bring up a box. Simply input the date and time of your observation, then look for the declination of Earth value. This the value of B. Once you have it, simply choose the right outline. Thus on 2013, April 2, at 0055UT, B = 18.7. Rounding up means we can take B = 19, and so we would use a Saturn outline of +19° for a Saturn sketch at that date and time. All of the Saturn blanks were drawn by David Gray.

P.G. Abel, *Visual Lunar and Planetary Astronomy*, The Patrick Moore
Practical Astronomy Series, DOI 10.1007/978-1-4614-7019-9,
© Springer Science+Business Media New York 2013

Jupiter Blank

Mercury/Venus/Mars/Uranus Planetary Blank

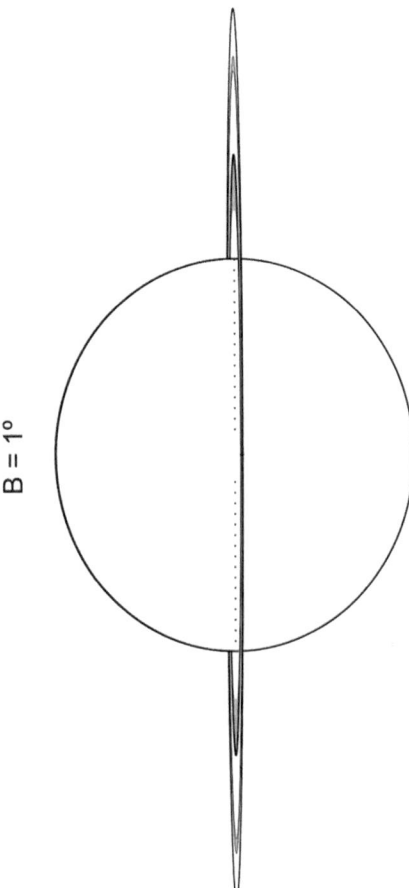

B = 1°

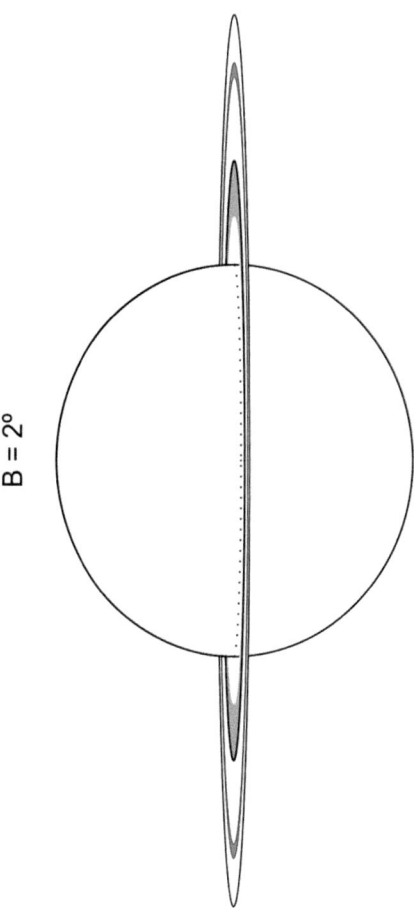

B = 2°

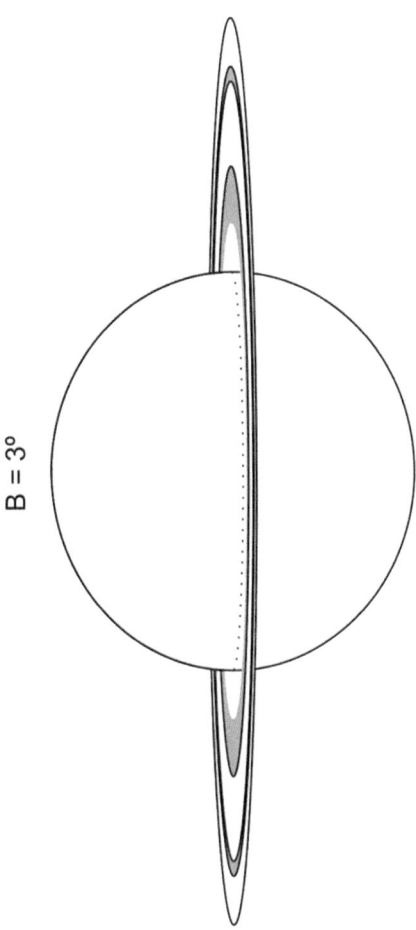

B = 3°

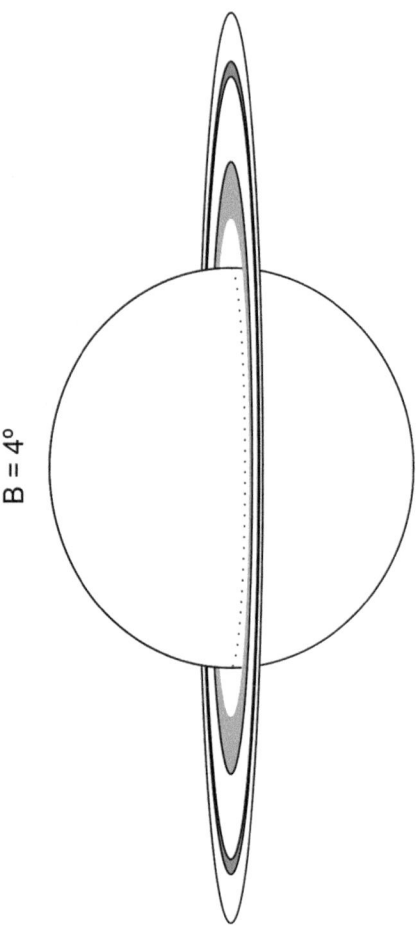

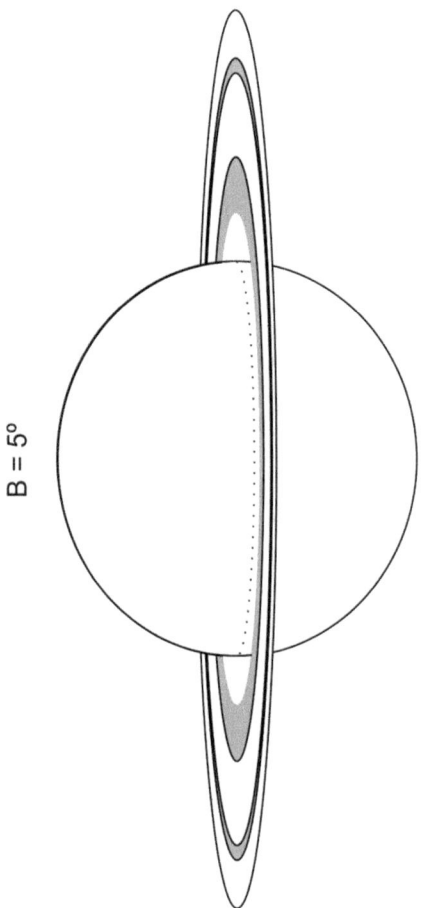

B = 5°

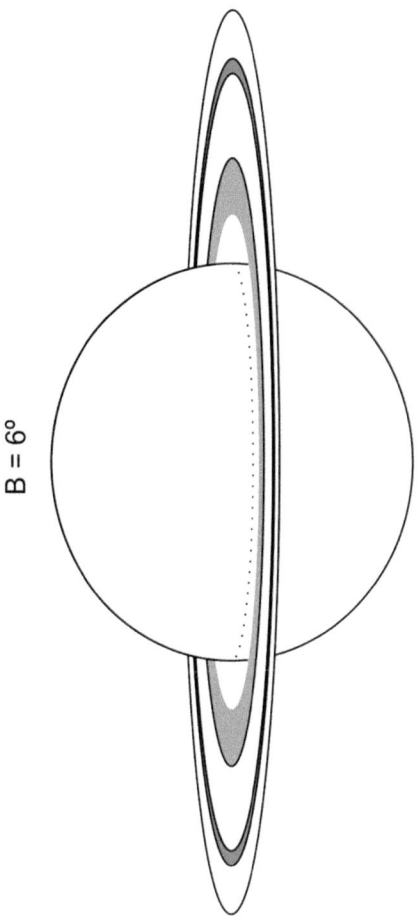

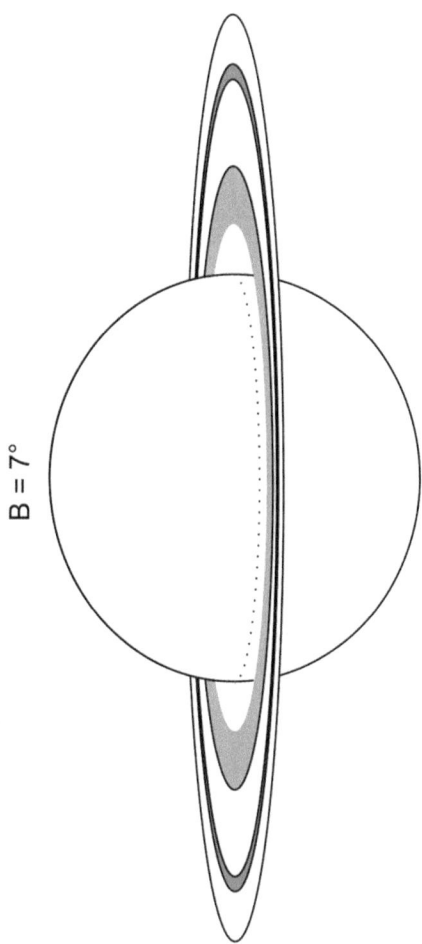

B = 7°

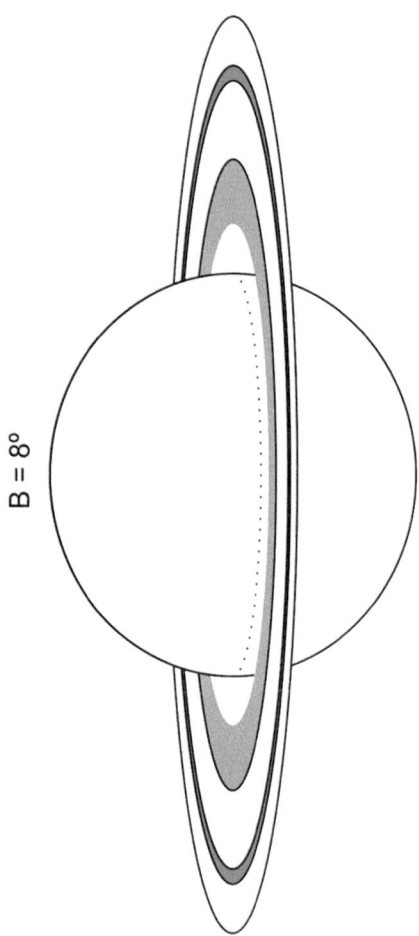

B = 8°

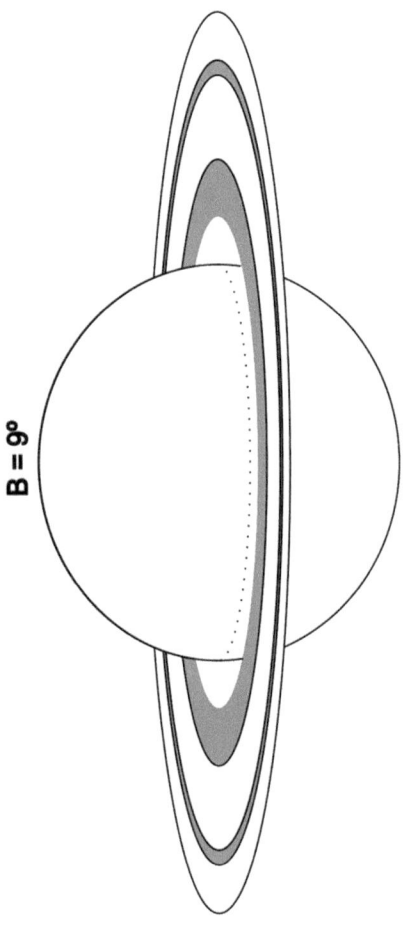

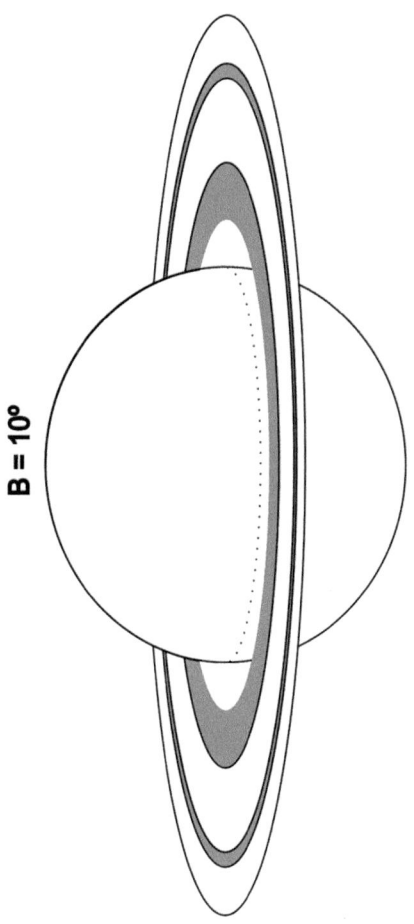

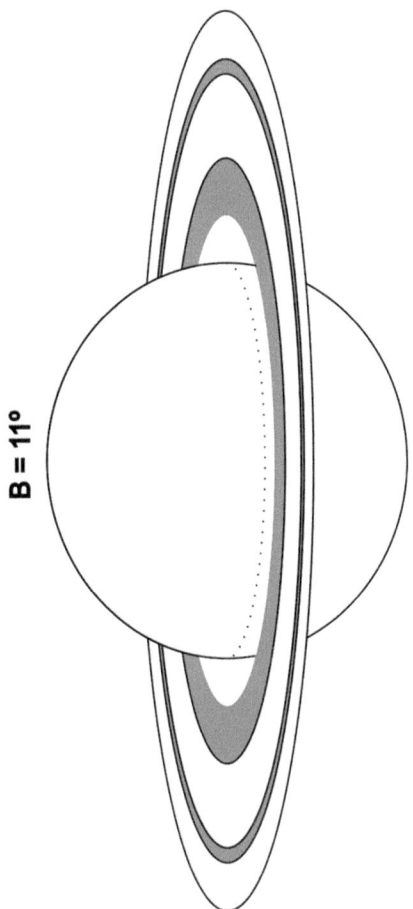

B = 11°

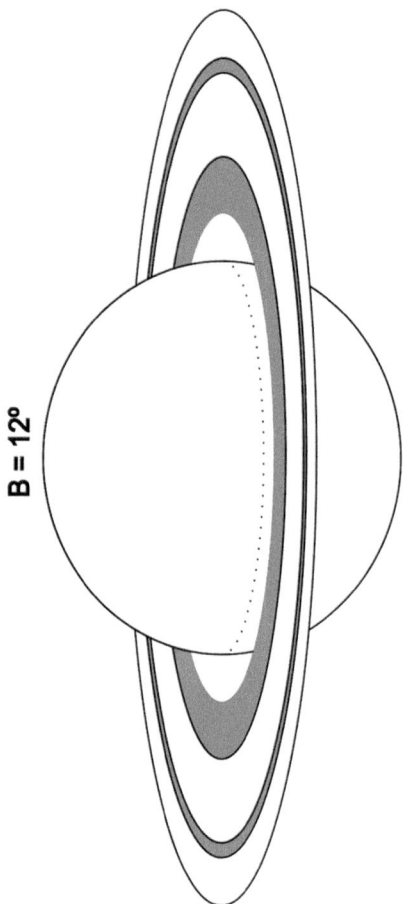

B = 12°

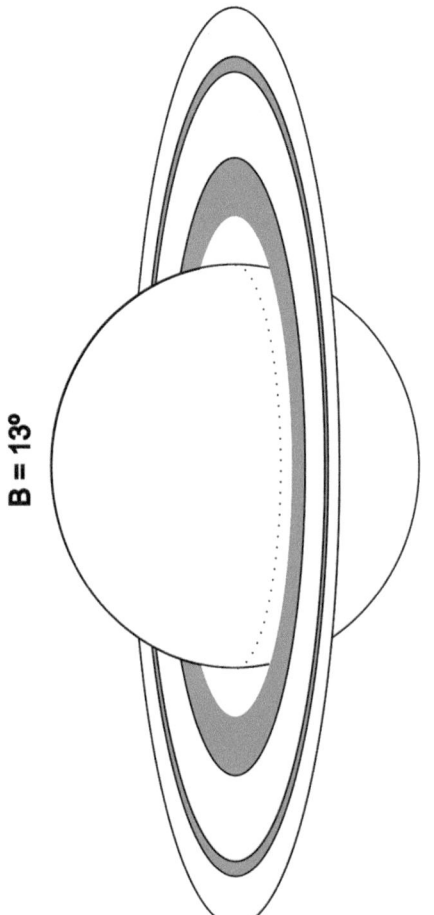

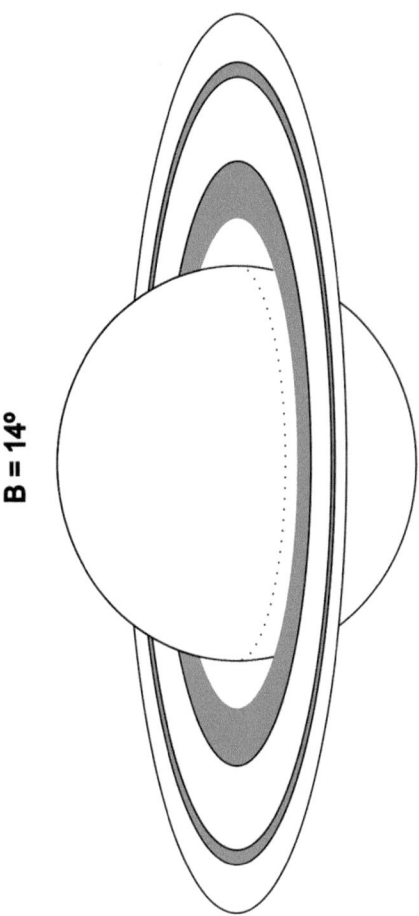

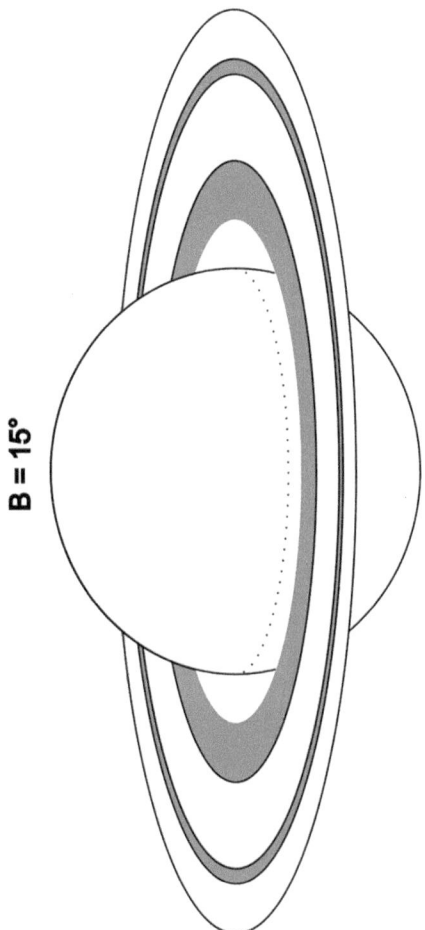

B = 15°

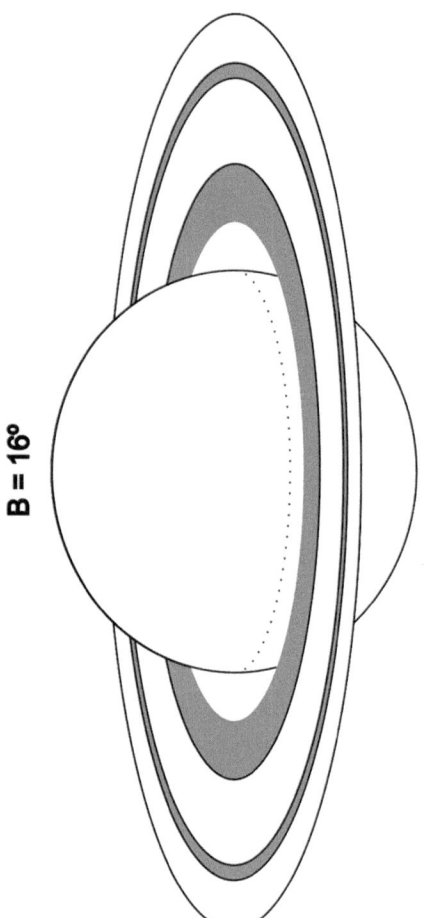

B = 16°

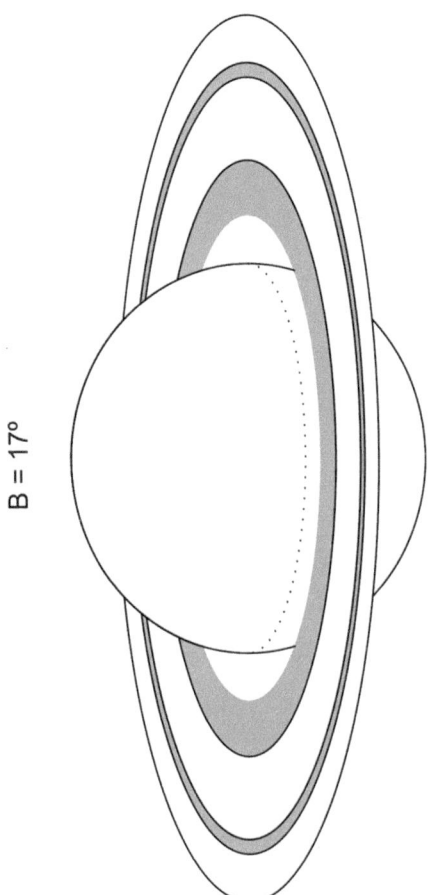

B = 17°

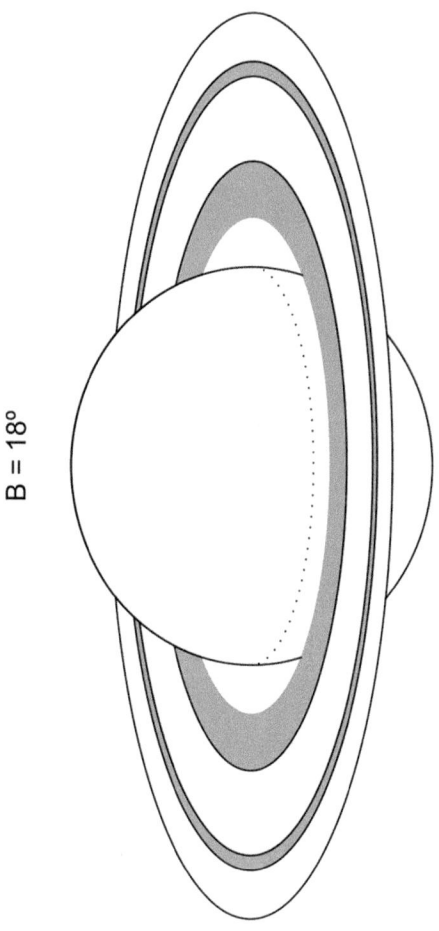

B = 18°

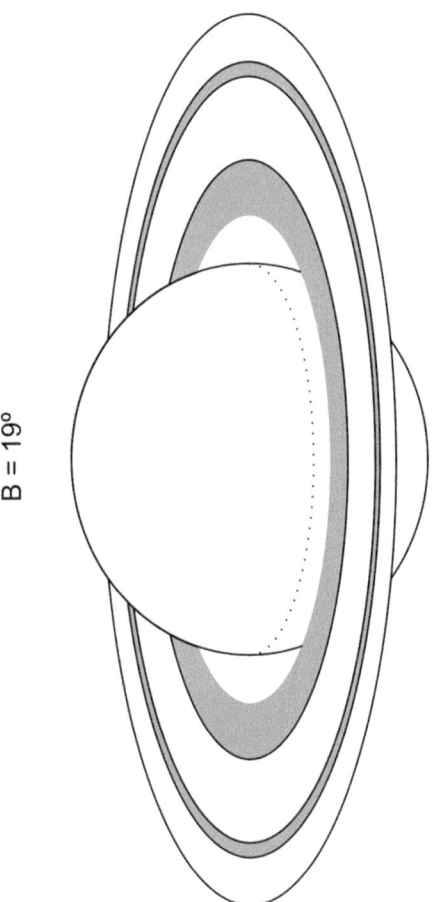

B = 19°

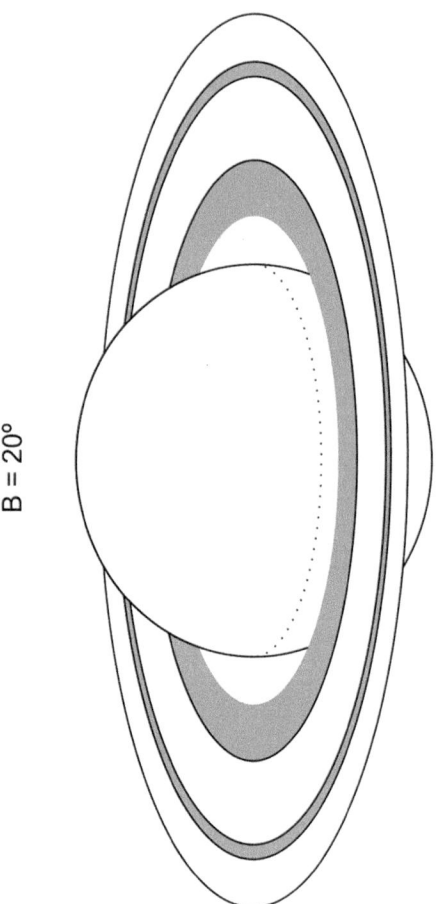

B = 20°

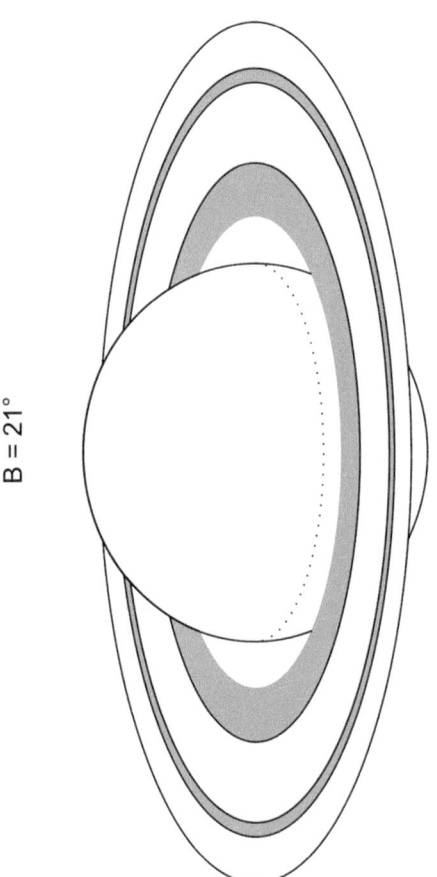

B = 21°

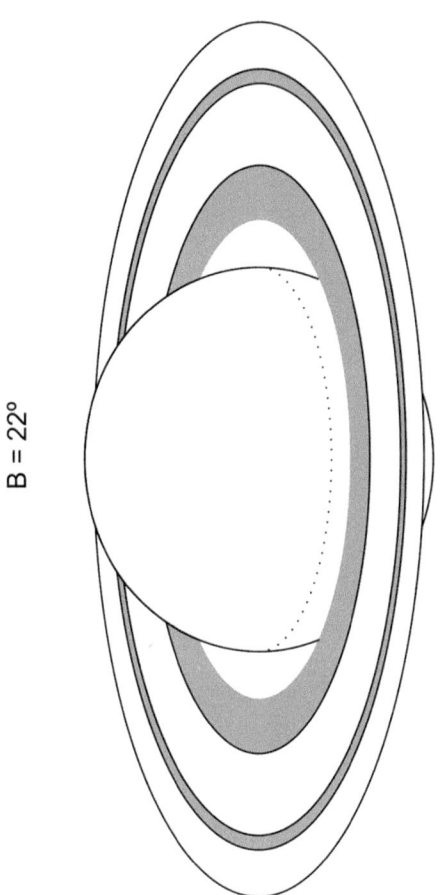

B = 22°

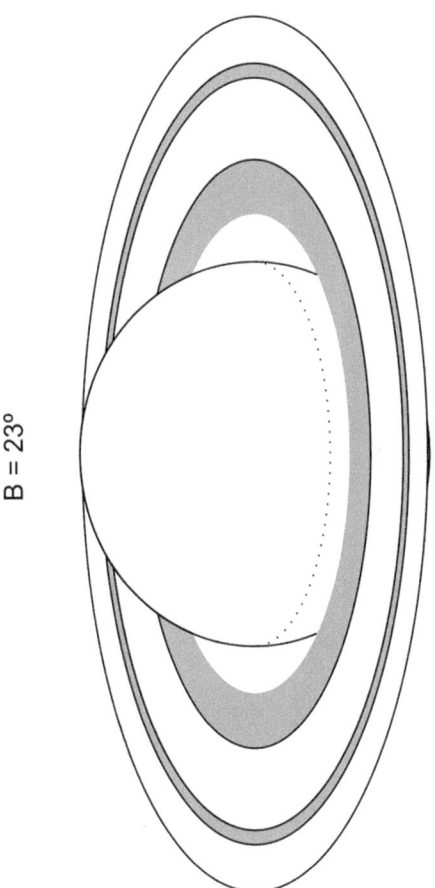

B = 23°

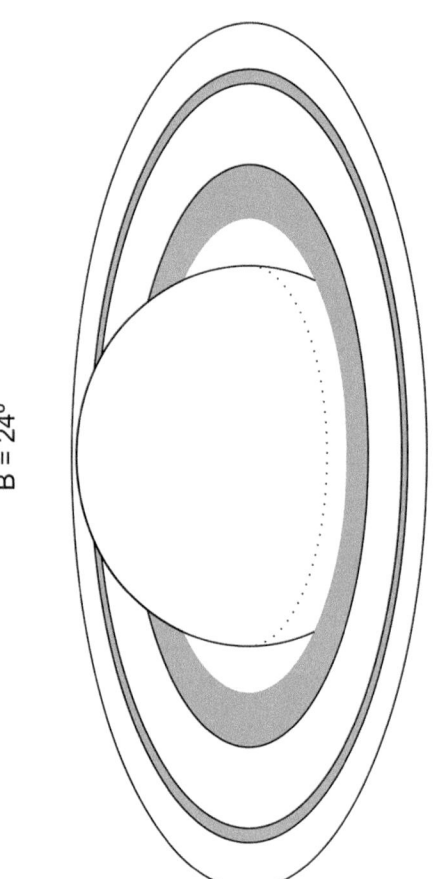

B = 24°

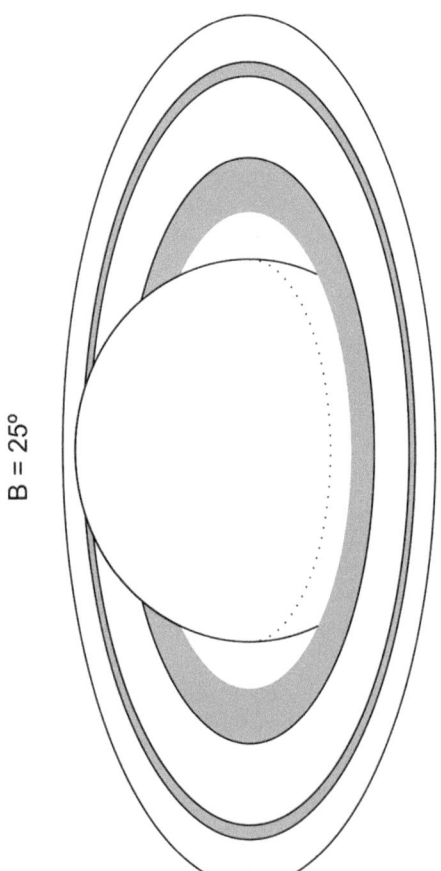

B = 25°

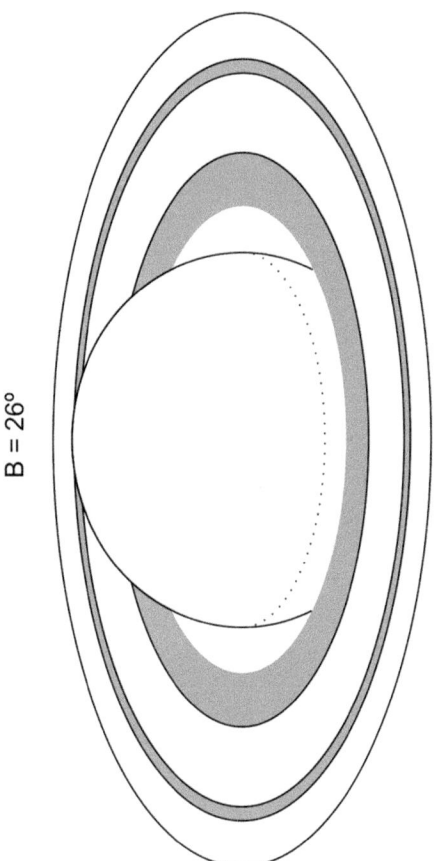

B = 26°

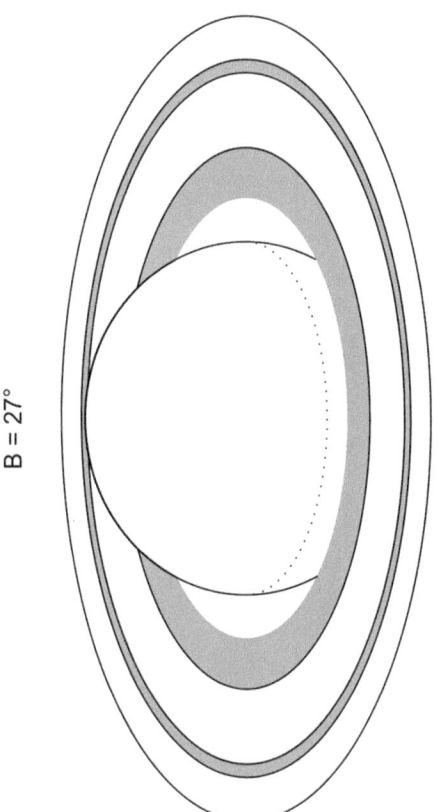

B = 27°

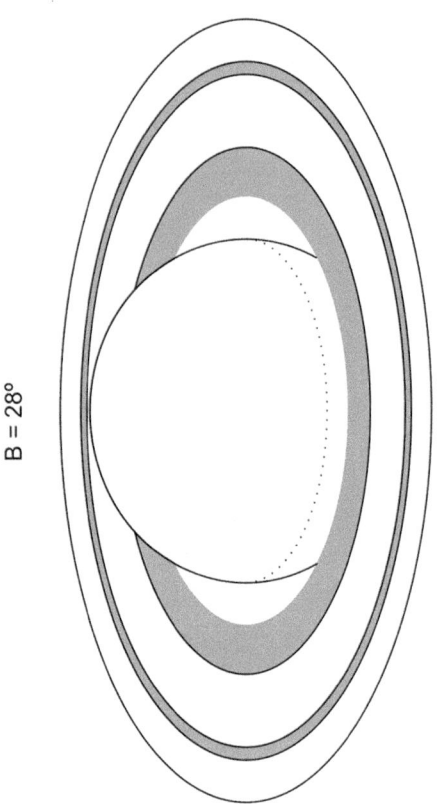

B = 28°

a

Index

P.G. Abel, *Visual Lunar and Planetary Astronomy*, The Patrick Moore
Practical Astronomy Series, DOI 10.1007/978-1-4614-7019-9,
© Springer Science+Business Media New York 2013